Design Guidelines for Surface Mount and Fine-Pitch Technology

Other books in the Electronic Packaging and Interconnection Series

Related books of interest

Design Guidelines for Surface Mount and Fine Pitch Technology

second edition

Vern Solberg

McGraw-Hill

New York San Francisco Washington, D.C. Auckland Bogotá
Caracas Lisbon London Madrid Mexico City Milan
Montreal New Delhi San Juan Singapore
Sydney Tokyo Toronto

McGraw-Hill

A Division of The McGraw·Hill Companies

hc   1 2 3 4 5 6 7 8 9 FGR/FGR 9 0 0 9 8 7 6 5

Library of Congress Cataloging-in-Publication Data
Solberg, Vern.
    Design guidelines for surface mount and fine-pitch technology / by
Vern Solberg. — 2nd ed.
       p.   cm.
    Includes index.
    ISBN 0-07-059577-1
    1. Printed circuits—Design and construction.   2. Surface mount
  technology.   3. Fine pitch technology.   I. Title.
  TK7868.P7.S637 1995   95-17264
  621.3815'31—dc 20   CIP

Acquisitions editor: Steve Chapman
Editorial team: Joanne Slike, Executive Editor
                Andrew Yoder, Managing Editor
                Melanie Holscher, Book Editor
Production team: Katherine G. Brown, Director
                Susan E. Hansford, Coding
                Brenda S. Wilhide, Computer Artist
                Wanda S. Ditch, Desktop Operator
                Linda L. King, Proofreading
                Joann Woy, Indexer
Design team: Jaclyn J. Boone, Designer
             Katherine Stefanski, Associate Designer

0595771
EL3

Contents

3 Land pattern development for SMT   *47*

6 SMT layout and guidelines for flexible circuits *135*

Preface

SEVERAL FACTORS HAVE CONTRIBUTED TO THE EVOLUTION of surface mount technology but component complexity and assembly auto-mation are generally recognized as the most significant. *Surface Mount Technology* has evolved from an alternative assembly process to what is currently established as the primary method of manufacturing electronic products. The original motivation for using the methodology was to meet the demand for product miniaturization, to combine functions of several circuit boards onto one, or simply the need to reduce overall system size. To accommodate the industries demand for a smaller size-to-function ratio for electronics, package technology for integrated circuits continues to move toward more complex solutions, specifically, fine-pitch and array packaging. Integrated circuits developed originally for single functions, although still widely used for a number of applications, have evolved to multiple function devices with higher I/O, and closer lead spacing. In parallel, the evolution of passive or discrete function device families has moved toward a smaller physical outline.

To benefit from the advances in component packaging technology, the circuit board designer must fully understand the requirements for assembly processing. Land pattern geometry, circuit board material, surface finish, and fabrication methods can significantly influence both product cost and manufacturing efficiency. Designers developing the circuit structure, for example, might not be closely aligned to assembly operations. It is not uncommon to find talented designers attempting to develop a potentially high-volume SMT product with little knowledge of assembly methods or attachment options. Many are not working within the manufacturing environment and have few sources for guidance when preparing a design for automation.

Standards for both devices and solder attachment criteria are well established by the industry. Devices are documented through national organizations such as the EIA/JEDEC in the U.S. and EIAJ

(Japan), while solder joint requirements are defined by the industry in J-STD-001. The requirements are based on specific use categories of a product to meet established goals for reliable operation over the product's life. It is the responsibility of the designer to learn the relationship of these elements during the planning stage of the surface mount product giving the assembly process specialist the opportunity to meet the attachment criteria. In preparation for surface mount technology, the designer should understand printed circuit fabrication methods, materials, and supplier capability. In addition, understanding each phase of the assembly process is vital. This understanding will include physical features and specific data required for developing tooling or implementing automated assembly technology.

The information and guidelines in this book have been developed through several years of practical use in the industry. Although the designer might find some differences in practice or methods for assembly, he or she should always strive to furnish a design that achieves the end-use goals for performance while ensuring manufacturing efficiency and product reliability.

Introduction

DESIGN OF THE PRINTED CIRCUIT BOARD FOR SURFACE mount technology is an interactive process that must weigh several predefined elements of materials and manufacturing methods. The circuit structure, either rigid or flexible, must accommodate recognized fabrication tolerance limits while providing the elements needed for several assembly process methods.

Design Guidelines for Surface Mount and Fine Pitch Technology was written to prepare the circuit board designer, engineer, or CAD specialist with many of the proven techniques needed for developing a reliable assembly process-compatible product. The book emphasizes why today's professional must be cognizant of the relationship between substrate design and the manufacturing process, because the design often influences the ultimate success of the process.

The text and illustrations provide you with step-by-step procedures in developing the most cost-effective product possible using surface mount and fine-pitch technology. Also defined are details related to device standards, alternative packaging methods, and circuit fabrication options. Written by a designer for designers, the detailed information is useful to several disciplines, including printed circuit designers, manufacturing engineers, as well as anyone developing electronic circuit assemblies.

Because of circuit complexity and fabrication methods in use today for surface mount applications, CAD-generated data and laser-developed photo-tools have become commonplace. A successful surface mount assembly is possible only when the circuit design meets the criteria for process efficiency and reliability. *Design Guidelines for Surface Mount and Fine Pitch Technology* is a workbook not an overview. Each chapter is written, illustrated, and detailed to guide the user in the implementation of the technology with process-proven techniques for SMT rather than untried or unproven theories.

This data can be efficiently utilized in several segments of manufacturing. CAD data can be modified for use in the manufacture of the printed circuit structure, bare-board test fixturing, and multiple tasks related to the assembly process. The land pattern data, for example, is used to prepare the solder paste stencil, the component data for automated placement programming, and physical aspects are used for other fixtures needed in the process, including the profile routing of boards furnished in a panel format.

Electrical and physical CAD data can be employed for testing the finished assembly as well. Automated testing of surface mount assemblies are efficiently performed by specialized computer systems. Electrical models are developed in the test system program for each device on the board and accessed through test sites (via holes or land patterns). The test site coordinates are combined with the electrical schematic to identify interconnecting points between contact for rapid circuit analysis.

xvi

Planning for
surface mount design

SURFACE MOUNT TECHNOLOGY (SMT) IS A MANUFACTURING process that, through miniaturization, provides increased component density for electronic products. This process offers the user a means of producing electronic products in high volume with a minimum of labor through robotic assembly.

Components, assembly equipment, and materials for attaching the surface mount devices continue to improve, meeting the exceedingly higher demands for quality and quantity. To support this technology all segments of the industry have striven to standardize in areas that complement each other. Industry-sponsored organizations and government agencies generally work together to review and approve standards for quality and/or reliability. When a component manufacturer introduces a new package to the industry it is usually a product that benefits the user and is a welcome addition to the complement of devices already in use. The component industry also will develop packaging that can be used or adapted to a wide variety of assembly systems.

Newly developed assembly equipment having different functions in the overall SMT process is made compatible through the cooperative effort of many standards organizations. For example, the working zones of different brands of machinery are at or near the same height from the floor and will have compatible conveyer systems with material flow and width adjustment from a common direction. This compatibility allows the systems engineer to choose the best combination of equipment for product assembly without compromising efficiency or precious resources.

With the use of JEDEC (Joint Electronic Device Engineering Council), EIA (Electronic Industry Association), and EIAJ (Electronic Industry Association of Japan), standard package types generally will be a safe rule. Component buyers now have more choices of suppliers. The standard device packaging is compatible with the assembly systems and when the land patterns are developed correctly, the result is a reliable cost-effective SMT assembly.

The planning phase for developing the electronic product with surface mount technology should include the review of all significant issues. Issues to be defined, which impact the product's manufacturability are performance, reliability and overall quality, and safety. In general, the designer of the surface mount PC board is given the opportunity to transform an idea from a nontangible electronic schematic diagram to a physical, functional unit of usable hardware. When the design of the board is completed, the end result of the design specialist's efforts will be expected to achieve standards of excellence in every conceivable category:

☐ Design for Manufacture (DFM)

☐ Design for Assembly (DFA)

☐ Design for Test (DFT)

☐ Design for Performance (DFP)

☐ Design for Reliability (DFR)

☐ Design for Safety (DFS)

☐ Design for Recycling (DFR)

The last item listed above is a relatively new issue for the electronic industry. But with the millions of tons of electronics that eventually reach obsolescence each year, the developer or seller will be under increasing pressure to take responsibility for the products demise.

In reality, the planning and development of the product should be the cognizant responsibility of individuals representing each of these specialized areas. As a team, the designer becomes the key player to implement a Design for Excellence effort that will include the respective experts in all the areas defined. The product's development will flow much smoother if all the critical issues are understood and defined at the start. As a basic planning strategy, first define performance criteria, then determine specific industry standards or specifications the product must meet, and finally establish the working environment in which the product is expected to operate or survive.

Initially, the development team must understand the category of product that is to be developed. The industry has helped define these product categories by identifying expectations of performance. In the United States, The Institute for Interconnecting and Packaging Electronic Circuits (IPC) membership identified three basic "Use Categories" defined as Class 1, Class 2, and Class 3 products. These categories include: General Electronic Products, Dedicated Service Electronic Products, and High Reliability Electronic Products.

These basic category definitions have been recognized by the international community as well. The IEC (International Electrotechnical Commission) members have agreed on the same basic categories of products, however, the reference to "Class" was not acceptable because it might imply a substandard product. The IEC agreed on the three categories, but preferred the terms Level A, Level B, and Level C as a universal definition of product categories.

The design of the printed circuit board influences the performance of the product to an extent, but basic material selection, specifically component devices, substrate selection, and the assembly process can determine the level of compliance. To clarify the Use Categories of electronic products further, the following explanation is offered for consideration.

General electronic products

This performance category includes consumer products, some computer and computer peripherals, as well as general-use products suitable for applications where cosmetic imperfections are not as important and the major requirement is a function of the printed circuit board or assembly (IPC Class 1 or IEC Level A).

Dedicated service electronic products

Equipment defined as a dedicated service product includes communication equipment, sophisticated business machines, instrumentation, and also equipment where high performance and extended life are required and uninterrupted service is desired, but not critical. In this category, minor cosmetic imperfections are allowed (IPC Class 2 or IEC level B).

High-reliability electronic products

This category might include high-grade commercial products, but generally it relates to equipment where continued or uninterrupted service on demand is critical. In this category equipment downtime cannot be tolerated and products must function when required—such as life support, emergency systems, and products that ensure vital services in emergency or severe environments, (IPC Class 3 or IEC Level C).

The compliance to the above categories or quality level is assessed using approved company or industry standards. In the United States, and also several countries without a large standards organization, companies recognize specific IPC (Institute for Interconnecting and Packaging Electronic Circuits), and ANSI/IPC (The American National Standards Institute) standards or guidelines to evaluate the quality of the electronic product. Although some companies have developed or prefer to develop in-house standards and specifications for design and manufacturing the references, guidelines and recommendations furnished in this manual reflect standards or requirements defined by recognized industry sources. Other standards can be applied to a product to meet specific requirements as referenced in Table 1-1.

■ Table 1-1

U.S. industry	EIA	IPC	ASTM	JEDEC	IEEE	ASME
	Private sector				Government	
U.S. national	ANSI				NIST	
	American National Standards Institute				National Institute STD & TECH	
Military	DOD		MOD		Other groups	
	U.S.		U.K.		NATO/...	
International	ISO			IEC		
European	CEN CENELEC			EC 92		
	Private sector				Government	
Asian	JPCA	JTPIA	JSI		MITI	

Some of the standards developed in the United States are adapted by other national and international organizations. These standards might differ to an extent because of the specific application, or in the case of the IEC or ISO standards, because of translation re-

4

quirements. Table 1-2 illustrates some examples of ANSI/IPC documentation that is or will be adapted for use in either United States Government specifications or as an International standard.

■ Table 1-2

Topic	Industry	Military	International
Terms	IPC-T-50	IPC-T-50	IEC-194
Design	IPC-D-275	IPC-D-275	IEC-326-3
	IPC-D-949		
	IPC-D-319		
	IPC-D-249	MIL-STD-2118	
	IPC-SM-782		
Materials	IPC-L-108	MIL-P-13949	IEC-249
(Rigid)	IPC-L-109		
	IPC-L-112		–3, –4, –5, –6
	IPC-L-115		
	IPC-L-125		–7, –8, –9, –10
(Flex)	IPC-FC-231	IPC-FC-231	–11, –12, –13, –14
	IPC-FC-232	IPC-FC-232	
	IPC-FC-233	IPC-FC-233	–15, –16, –17, –18
	IPC-FC-241	IPC-FC-241	
Performance	IPC-RB-276	MIL-P-55110	IEC-326
	IPC-ML-950		
	IPC-SD-320	MIL-P-50884	–4, –5, –6, –7
	IPC-FC-250		
	IPC-HM-860	MIL-P-RRRRR	–8, –9, –10, –11
Assembly	IPC-SM-780	MIL-P-28809	IEC-321
	IPC-SM-770	MIL-STD-454	
	IPC-S-815	MIL-STD-2000	–1
	ANSI/J-STD-001		
Acceptability	IPC-A-600	~	ISO-9000
	IPC-A-610		ISO-9001
	IPC-SS-605		ISO-9002
	IPC-SS-615		ISO-9003
Process Control	IPC-PC-90	IPC-PC-90	

Specific printed circuit board and surface mount design-related standards are available to assist the board designer. The IPC committees that developed these standards are professionals working in the field who supply these documents so that users can benefit from their collective knowledge and experience.

Three standards that apply directly to the design of printed circuit boards and surface mount are:

1. ANSI/IPC-D-275, Design Standards for Rigid Printed Circuit Boards and Rigid Printed Board Assemblies.

2. The IPC-SM-782, Design and Land Pattern Standards for SMT and IPC-D-279.

3. Reliability Design Guidelines for Surface Mount Technology Printed Board Assemblies.

Other related IPC documents that can be referenced for other elements of electronic materials and manufacturing are listed in Table 1-3.

■ Table 1-3

IPC-T-50	Terms and definitions
IPC-SC-60	Post Solder Solvent Cleaning Handbook
IPC-AC-62	Post Solder Aqueous Cleaning Handbook
IPC-CM-78	Guidelines for surface mounting
IPC-D-249	Design standard for flexible single- and double-sided printed boards
IPC-D-319	Design standard for rigid single- and double-sided printed boards
IPC-MC-323	Design standard for metal core printed boards
IPC-D-330	Printed wiring design guide
IPC-A-610	Acceptability of printed board assemblies
IPC-R-700	Rx for repair and modification of printed wiring boards
IPC-SM-780	Guidelines for component packaging and interconnection with emphasis on surface mounting
IPC-S-804	Solderability test methods for printed wiring boards
IPC-S-805	Solderability tests for component leads and terminations
IPC-S-815	General requirements for soldering electrical connections and printed board assemblies
IPC-AJ-820	Assembly and Joining Handbook
IPC-CC-830	Electrical insulating compounds for printed board assemblies
IPC-SM-840	Qualification and performance of permanent polymer coating (solder mask) for printed boards
IPC-ML-949	Design standard for rigid multilayer, printed boards

Defining the environmental use category

The "Use Category" of the product is most often determined by the environment or how and where the assembly or system is ex-

■ Table 1-4

Use category	Worst use environment						Accelerated testing			
	Tmin °C	Tmax °C	ΔT1) °C	tD hrs	cycles/ year	years of service	Tmin °C	Tmax °C	ΔT1) °C	tD min
1 Consumer	0	+60	35	12	365	1-3	+25	+100	75	15
2 Computers	+15	+60	20	2	1460	~5	+25	+100	75	15
3 Telecomm	−40	+85	35	12	365	7-20	+25	+100	75	15
4 Commercial aircraft	−55	+95	20	2	3000	~10	0	+100	100	15
5 Industrial & automotive passenger compartment	−55	+65	20 & 40 & 60 & 80	12 12 12 12	185 100 60 20	~10	0	+100	100	15
							& "COLD"3)			
6 Military ground & ship	−55	+95	40 & 60	12 12	100 265	~5	0	+100	100	15
							& "COLD"3)			
7 Space leo geo	−40	+85	35	1 12	8760 365	5-20	0	+100	100	15
							& "COLD"3)			
8 Military avionics a b c	−55	+95	40 60 80 & 20	2 2 2 1	500 500 500 1000	~5	0	+100	100	15
							& "COLD"3)			
9 Automotive under hood	−55	+125	60 & 100 & 140	1 1 2	1000 300 40	~5	0	+100	100	15
							& "COLD"3) & LARGE ΔT"4)			

pected to operate or survive. In establishing product use categories the industry has identified nine typical environments related to thermal conditions in which the product must operate as well as the expected service life cycle as listed in Table 1-4.

Although products can fall into different use categories, the design of each product utilizes common guidelines and design rules. Material and device selection might differ a great deal based on the product's operating environments. As an example, high-performance laminates and a better grade of copper are required to meet small via hole or finer lines and spaces on higher density circuit boards. Other systems might require high-performance base materials that have a higher glass transition point or improved dielectric constants because of the extreme operating temperature.

Devices have operational temperature limits as well. Most commercial electronic component families have functional limits generally defined by the material used in manufacturing. As an example, surface mount resistors, capacitors, and ICs made with ceramic materials are rated to operate at far higher temperature limits than those packaged using plastics.

Standards and guidelines are developed to assist engineers in the selection of materials and components. A typical document developed to assist engineers select and test materials, as well, as devices is the *IPC/SM/785—Guidelines for Accelerated Reliability Testing of Surface Mount Attachments*.

Defining SMT assembly type and complexity

The printed circuit board designer must have a basic understanding of the assembly processes used to fabricate and assemble the surface mount product. Because preferred devices are designed for surface mount attachment, components can be mounted on either the primary, secondary, or both sides of the circuit board. To clarify and define all assembly options, the industry has agreed on a method of identifying the assembly types as related to complexity. See Table 1-5.

The assembly sequence and solder methods used to process each of these assembly types is unique. The designer must define the limits of the equipment and technical capability available for assembly before the design program is started.

The details illustrated in Fig. 1-1 compare the most common assembly types just noted. For example, if a process is limited to only wave solder attachment, the surface mount devices are typically attached on the secondary side (Type 1A or 2C—Single assembly).

■ **Table 1-5 Operating temperature and expected life cycle.**

Product category	Min. temp.	Max. temp.	Yrs of serv.
Consumer	0	+60	1–3
Computer	+15	+60	>5
Telecom	–40	+85	7–20

■ Table 1-5 Continued.

Product category	Min. temp.	Max. temp.	Yrs of serv.
Aircraft	–50	+95	5–10
Indust/auto	–55	+65	>10
Underhood	–55	+125	5–6
Military	–55	+95	>5
Space	–40	+85	5–20

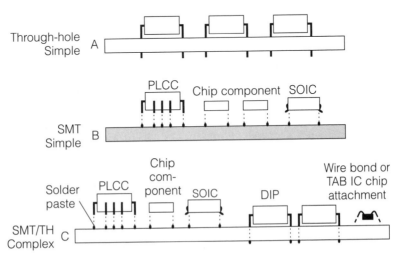

Through-hole Simple A

SMT Simple B
PLCC   Chip component   SOIC

SMT/TH Complex C
Solder paste   PLCC   Chip component   SOIC   DIP   Wire bond or TAB IC chip attachment

■ **1-1** *Although complexity of the circuit assembly may vary to a degree, IPC has defined the surface mount assembly with devices on only one side as Assembly Type 1.*

On the other hand, if both wave solder and reflow soldering systems are available for assembly processing, it will be possible to develop assembly capability for each of the assembly types described, simple or complex.

Even though a great majority of components are available in surface mount packages it is often necessary to mix various surface mount package types with the leaded pin-in-hole (PIH) devices on the same substrate. The placement and interconnection of these dissimilar components is in itself a major challenge for the designer of the PC board.

Adapting SIP, SIM, and MCM

Using miniature surface mount devices has provided layout flexibility to a point, but by adapting the popular SIP (Single Inline Pin) and SIM (Single Inline Module) configuration, as shown in Fig. 1-2, one or more complex SMT circuits can be stacked on edge.

Type 2

A 2-Sided through hole (not recommended)

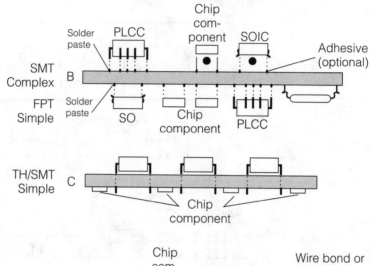

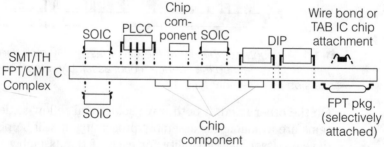

■ **1-2** *Assemblies with devices mounted on both primary and secondary sides of the circuit board are classified as Assembly Type 2 with varying degrees of complexity.*

Companies attempting to relieve the dilemma of ever-decreasing space, might combine the function of several integrated circuits into one or two multiple function custom devices. This module partitioning of multiple functions in a single application-specific integrated circuit (ASIC) technique can alleviate PC board gridlock, thereby allowing the designer to use fewer circuit layers.

Planning the SMT assembly

There is no one solution to everyone's packaging requirements, but through careful planning, assemblies will meet the goals of quality, reliability, and cost. Planning for the most economical and reliable SMT assembly requires attention to the following factors:

☐ Use of standard devices.

☐ Understanding of layout conventions.

☐ Adapting standard land pattern geometry.

☐ Providing adequate device spacing.

☐ Designing for automation.

Component package selection

Component selection for surface mount starts with a review of the manufacturers and distributors. Catalogs and directories are published each year listing thousands of devices available in surface mount packages. Use these sources as a starting point, and check the availability of the device before starting the PC board design. Many of the components will be on the shelf; however, what has happened on several occasions is that the device announced for availability on a given date is either not yet ready for distribution or not available in the volume you require. The other factor that impacts the start-up of a product is the shortage of allocated components throughout the industry.

The designer of the PC board must have a reasonable understanding of all materials and processes related to manufacturing the surface mount product. In the following chapters, each of the issues that impact the success of the surface mount product are presented in a logical order. The information, although explained and illustrated in detail, is simple and to the point. It is my goal to furnish the designer with the tools and overall knowledge of all aspects of the technology. Having defined the critical issues related to materials, the assembly process, the products function, and performance specifications, the designer of the PC board will be better prepared to achieve the level of excellence expected.

Component selection
for SMT

Components for surface mount technology

COMPONENT RESEARCH FOR SURFACE MOUNT TECHNOLOGY
starts with a review of suppliers, manufacturers, and distributors.
Catalogs and directories published each year list thousands of devices available in surface mount packages. Use these sources with
caution and check the actual availability of the device before starting the PC board design. Many general use components are on the
shelf; however, some of the new devices announced for availability
on a given date are not always ready for distribution or available in
the volume you require. Another factor that impacts the start-up
of a product is the shortage of components in high demand
throughout the industry.

The use of JEDEC, EIA, and EIAJ standard package types is generally a safe rule. Component buyers will have a better choice of
suppliers, and the devices are more likely to be compatible with
the assembly systems, solder process, and land patterns provided
on the PC board.

Mixed technology

It might be necessary to mix various SMT package types with the
leaded pin-in-hole (PIH) devices on the same substrate. The placement and interconnection of these dissimilar components are a
major challenge for the designer of the PC board. Although PIH
parts can be modified for surface attachment, the designer should
attempt to locate a surface mount alternative.

Planning component requirements

The material planning phase of the development has a critical impact on the manufacturing efficiency at each level of the assembly process. There might be no single solution to meet the packaging need of specific requirements, but through careful selection the surface mount assembly will provide a reliable, cost-effective product. Planning for the most economical and reliable SMT assembly will require attention to the following factors:

☐ Selection of standard devices.

☐ Compatibility with the operating environment.

☐ Qualify reliable sources (2 or 3).

☐ Use standard packaging systems.

Standard devices for SMT

The component industry is very competitive worldwide. Many standards for devices have emerged because Brand X wanted a share of the market held by Brand Y. Although many SMT devices are simply a clone of another, most formal standards have evolved through industry agreement, established electronic industry associations, or standards organizations.

The United States-based EIA and JEDEC have documented the physical characteristics of most of the devices in general use. This includes several package types developed in Japan and Europe. The Japanese equivalent to the United States organizations is the EIAJ, and because a significant share of devices are supplied from Japan and other Asian countries, standards are often well established before they are introduced to the United States market.

One point that must be clear to the designer when adapting SMT to the printed circuit board assembly, is the need for accuracy in developing the attachment sites or land patterns. Surface mount packaging standards are established for most of the high-volume device families. In selecting the package configuration for a product, use suppliers of the standard configurations when possible. Avoid nonstandard devices unless the manufacturing source is very reliable.

Passive devices for SMT

Passive or discrete components are typically the most common, lowest cost, and most commonly used components on surface mount assemblies. Most devices in this category, for example, resistors and capacitors, have only two terminations while others might have three or more terminations depending on the specific function or application for which it has been designed. The devices shown in Fig. 2-1 are packaged in surface mount configurations for solder attachment directly to a PC board.

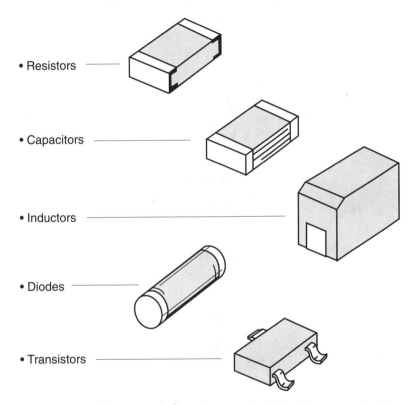

- Resistors
- Capacitors
- Inductors
- Diodes
- Transistors

■ **2-1** *Discrtete electronic devices are furnished in several unique shapes and sizes but all are equally compatible with solder attachment processes typical of SMT.*

Choosing passive components for the surface mount assembly often requires a great deal of research. When possible, select passive components that are uniform in type and size. Resistors and capacitors, for example, are available in the same dimensional shape. By selecting standard sizes, components will be available from several manufacturers at the most favorable prices.

Surface mount resistors are available in all standard values from several sources. As with capacitors, some sizes are more common than others. Higher wattage components will have a limited value selection. The 1206 size, and greater, is available with values clearly marked on the surface for easy inspection.

For the more general application, chip resistors use a thick film element on an alumina substrate. The end-cap terminations have a silver alloy base with tin-lead solder plating over a nickel barrier for compatibility with SMT assembly and solder systems. The resistor element is insulated with a resin or glass coating. For details, see Fig. 2-2.

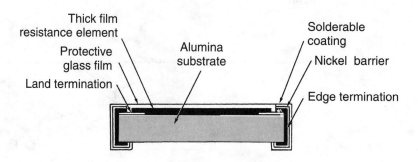

Standard chip resistor family

metric	inch
1005	0402
1608	0603
2012	0805
3216	1206
3225	1210
5025	2010*
6332	2512*

■ **2-2** *Chip resistors are furnished in several physical shapes and wattage ratings. The rectangular 1206 size ceramic-based device is considered a standard for ⅛-watt applications.*

A common chip size in use is the 1206 rated at ⅛ watt by most manufacturers. It is often substituted for ¼-watt applications with specific temperature range limits. The standard resistance decade is shown in Table 2-1.

Other sizes are available as well including the 1005/0402, 1608/0603, and 2012/0805 for miniature applications with low-watt ratings, and the 3225/1210 device for ¼ watt-applications. One-half-watt and 1-watt resistors are available in surface mount packages, but value selection is limited. Resistors are also available in the cylin-

C,D .25,.5	F 1.0	G 2	J 5	C,D .25,.5	F 1.0	G 2	J 5	C,D .25,.5	F 1.0	G 2	J 5
10.0	10.0	10.0	10.0	18.7	18.7	18.7	—	56.2	56.2	56.2	56.2
10.1	—	—	—	18.9	—	—	—	56.9	—	—	—
10.2	10.2	—	—	19.1	19.1	—	—	57.6	57.6	—	—
10.4	—	—	—	19.3	—	—	—	58.3	—	—	—
10.5	10.5	10.5	—	19.6	19.6	19.6	19.6	59.0	59.0	59.0	—
10.6	—	—	—	19.8	—	—	—	59.7	—	—	—
10.7	10.7	—	—	20.0	20.0	20.0	20.0	60.4	60.4	—	—
10.9	—	—	—	20.3	—	—	—	61.2	—	—	—
11.0	11.0	11.0	11.0	20.5	20.5	20.5	—	61.9	61.9	61.9	61.9
11.1	—	—	—	20.8	—	—	—	62.0	62.0	62.0	62.0
11.3	11.3	—	—	21.0	21.0	—	—	62.6	—	—	—
11.4	—	—	—	21.3	—	—	—	63.4	63.4	—	—
11.5	11.5	11.5	—	21.5	21.5	21.5	21.5	64.2	—	—	—
11.7	—	—	—	21.8	—	—	—	64.9	64.9	64.9	—
11.8	11.8	—	—	22.0	22.0	22.0	22.0	65.7	—	—	—
12.0	12.0	12.0	12.0	22.1	22.1	—	—	66.5	66.5	—	—
12.1	12.1	12.1	12.1	22.3	—	—	—	67.3	—	—	—
12.3	—	—	—	22.6	22.6	22.6	—	68.0	68.0	68.0	68.0
12.4	12.4	—	—	22.9	—	—	—	68.1	68.1	68.1	68.1
12.6	—	—	—	23.2	23.2	—	—	69.0	—	—	—
12.7	12.7	12.7	—	23.4	—	—	—	69.8	69.8	—	—
12.9	—	—	—	23.7	23.7	23.7	23.7	70.6	—	—	—
13.0	13.0	13.0	13.0	24.0	24.0	24.0	24.0	71.5	71.5	71.5	—
13.2	—	—	—	24.3	24.3	—	—	72.3	—	—	—
13.3	13.3	13.3	13.3	24.6	—	—	—	73.2	73.2	—	—
13.5	—	—	—	24.9	24.9	24.9	—	74.1	—	—	—
13.7	13.7	—	—	25.2	—	—	—	75.0	75.0	75.0	75.0
13.8	—	—	—	25.5	25.5	—	—	75.9	—	—	—
14.0	14.0	14.0	—	25.8	—	—	—	76.8	76.8	—	—
14.2	—	—	—	26.1	26.1	26.1	26.1	77.7	—	—	—
14.3	14.3	—	—	26.4	—	—	—	78.7	78.7	78.7	—

■ Table 2-1 Continued.

C, D .25, .5	F 1.0	G 2	J 5	C, D .25, .5	F 1.0	G 2	J 5	C, D .25, .5	F 1.0	G 2	J 5	C, D .25, .5	F 1.0	G 2	J 5
14.5	—	—	—	26.7	26.7	—	—	45.9	—	—	—	79.6	—	—	—
14.7	14.7	14.7	14.7	27.0	27.0	27.0	27.0	46.4	46.4	46.4	46.4	80.6	80.6	—	—
14.9	—	—	—	27.1	—	—	—	47.0	47.0	47.0	47.0	81.6	—	—	—
15.0	15.0	15.0	15.0	27.4	27.4	27.4	—	47.5	47.5	—	—	82.0	82.0	82.0	82.0
15.2	15.4	15.4	—	27.7	—	—	—	48.1	—	—	—	82.5	82.5	82.5	82.5
15.4	15.4	—	—	28.0	28.0	—	—	48.7	48.7	48.7	—	83.5	—	—	—
15.6	15.8	—	—	28.4	—	—	—	49.3	—	—	—	84.5	84.5	—	—
15.8	15.8	—	—	28.7	28.7	28.7	28.7	49.9	49.9	—	—	85.6	—	86.6	—
16.0	16.0	16.0	16.0	29.1	—	—	—	50.5	—	—	—	86.6	86.6	86.6	—
16.2	16.2	16.2	16.2	29.4	29.4	—	—	51.0	51.0	51.0	51.0	87.6	—	—	—
16.4	—	—	—	29.8	—	—	—	51.1	51.1	51.1	51.1	88.7	88.7	—	—
16.5	16.5	—	—	30.0	30.0	30.0	30.0	51.7	—	—	—	89.8	—	—	—
16.7	—	—	—	30.1	30.1	30.1	—	52.3	52.3	—	—	90.9	90.9	90.9	90.9
16.9	16.9	16.9	—	30.5	—	—	—	53.0	—	—	—	91.0	91.0	91.0	91.0
17.2	—	—	—	30.9	30.9	—	—	53.6	53.6	53.6	—	92.0	—	—	—
17.4	17.4	—	—	31.2	—	—	—	54.2	—	—	—	93.1	93.1	—	—
17.6	—	—	—	31.6	31.6	31.6	31.6	54.9	54.9	—	—	94.2	—	—	—
17.8	17.8	17.8	17.8	32.0	—	—	—	55.6	—	—	—	95.3	95.3	95.3	—
18.0	18.0	18.0	18.0	32.4	32.4	—	—	56.0	56.0	56.0	56.0	96.5	—	—	—
18.2	18.2	—	—	32.8	—	—	—					97.6	97.6	—	—
18.4	—	—	—									98.8	—	—	—

der shaped (MELF) configuration, with overall length and width dimensions similar to ceramic chip components.

The MELF is popular for wave-solder applications when the component is held to the secondary side of the substrate with epoxy. Verify the compatibility of the round component with the assembly equipment to be used before selecting the tubular-shaped body.

Packaging for automation

SMT components are supplied to the end user in one of three configurations: bulk, tube magazine, and tape-and-reel. Tape-and-reel packaging is generally preferable for automation, but for low-volume or prototype production, the tube magazine might be acceptable. Both tape-and-reel and magazine containers are clearly marked for efficient material control and they easily adapt to the automated SMT placement equipment. Loose packaging of chip type components is less desirable because the part must be handled with special feeders for the assembly equipment.

Chip resistors and capacitors are supplied from most sources on 8-mm and 10-mm tape-and-reel packaging. Each reel holds up to 4000 or 5000 chip parts. Component identification for material control is vital. The manufacturer's code number on the container is often the only way to verify contents, because not all chip component values are identified on the body surface.

Resistor networks

Resistor networks are available in many surface mount configurations from several major suppliers in SO-14, SO-16, and PLCC types; however, not all manufacturers conform to uniform standard. For example, companies manufacturing resistor networks use a thin ceramic material as a base. Each resistor element on the network array is applied using a resistive paste-like ink material that is fired or fused to the ceramic base at high temperature. (See Fig. 2-3.)

The overall size of these devices can vary somewhat because each manufacturer might use a slightly different process for packaging the ceramic resistor network. Resistor networks are available in an open leadless ceramic configuration as well as the leaded packaging previously noted.

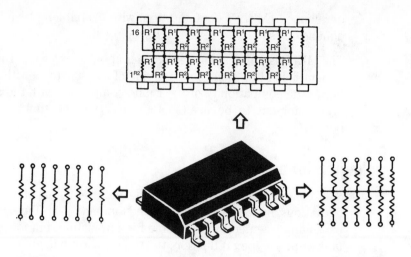

■ **2-3** *Surface mount resistor networks are available in many standard values. Custom networks are ordered directly from the manufacturer.*

A word of caution regarding the leadless resistor array: These unpackaged ceramic devices are thin and very brittle. If the designer is attaching the device to a laminated PC board material, ceramic devices larger than 5.0 mm (0.200 in) in length should be avoided.

Resistor networks are not controlled by the same uniform standards that apply to discrete device or IC manufacturers. When selecting resistor networks in the small outline (SO) configuration, the designer will find a significant difference from leading suppliers in both overall package width and thickness. Although the value of the device is the same from one manufacturer to the other, one might furnish a product in a standard narrow body SO type package while the other supplies the part in a medium, non-standard-width SO package. Review manufacturer's physical specifications carefully when choosing component sources. If the wrong package is designed into the product or an incorrect substitution is made when purchasing these devices, the company might suffer serious production delays.

Potentiometers for SMT

There is a multitude of single-turn cermet element potentiometers, commonly known as *Pots*, on the market for SMT PC boards. As the sources are researched, the designer will find that adjustable devices differ greatly from one manufacturer to another. It is important to choose a device that is packaged for automated assembly equipment, as well as being compatible with the high temperatures during solder-reflow processes, as shown in Fig. 2-4.

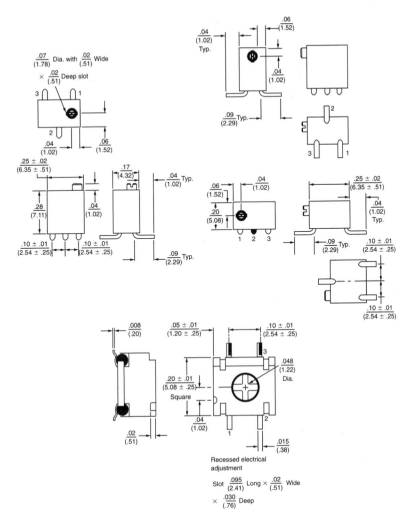

■ 2-4 *Miniature surface mount potentiometers are available from leading manufacturers however, few meet a specific standard outline or conform to a uniform contact pattern geometry.*

The devices shown in Fig. 2-5 are designed for flat- and right-angle mounting. Many manufacturers offer these devices with contacts on 1.27 (0.050 in) spacing with open or closed body construction, but they might have significant physical discrepancies between one supplier and another.

When mounting adjustable devices, a small amount of adhesive can be added to enhance the mechanical integrity of a part that requires frequent adjustment.

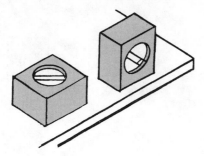

■ **2-5** *Potentiometer standardization is not as established as other component types. Choose devices that can be supplied from more than one source when possible.*

Monolithic capacitors

The selection process for monolithic capacitors is more than choosing the value and the dielectric. Additional choices include body size, end-cap termination, and material or plating.

Due to the continued miniaturization of hand-held electronics, such as games, communications equipment, and other consumer products, the component suppliers are asked to supply specific passive devices in smaller packages. The 1005/0402-size device is currently supplied for both resistor and capacitor applications, meeting the demands of these fast-growing, high-volume markets. Although this market segment is growing, assembly specialists continue to recommend the use of larger devices when possible. The smaller devices are difficult to handle with most automated pick-and-place systems, and the values available for capacitors are somewhat limited.

One of the more common capacitor sizes for values up to 0.10 uF is the 2012/0805 and 3216/1206. The 0.10-µF and the 0.15-µF capacitor is available in the 3216/1206 body, but the 3216/1210 offers a broader source selection to the designer (see Fig. 2-6). The narrower 3216/1206 component will mount on the 3225/1210 land pattern. Higher voltage and capacitance values are available in the larger size with several dielectric options.

The physical size of capacitors are similar to the resistor package family as shown in Fig. 2-7, but they are not exactly the same overall. Refer to EIA or manufacturer's standards to verify dimensions and tolerances.

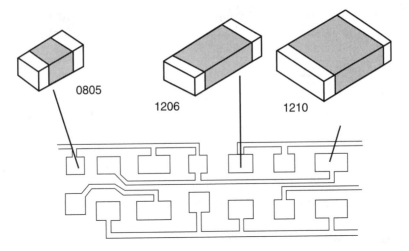

2-6 *Value and dielectric requirements might force the use of several capacitor sizes on the same assembly. Changing from a lower to a higher value might require a modification of the substrate in order to provide the correct land pattern.*

Standard chip capacitor family

metric	inch
1005	0402
1310	0504
1608	0603
2012	0805
3216	1206
3225	1210
4332	1812
4564	1825*

2-7 *The chip capacitor family is available in several standard sizes accommodationg a wide range of value and operating voltages.*

Chip capacitor selection

Dielectric type

The choice of dielectric is largely determined by the requirement of the temperature stability.

COG

The COG is an ultra stable Class I dielectric, with negligible dependence of electrical properties on temperature, voltage, frequency, or time. It is used in circuits requiring stable performance.

X7R

The X7R is a stable Class II dielectric with predictable change of properties with temperature, voltage, frequency, and time. It is used as blocking coupling, by-pass, and frequency discriminating elements. This dielectric is ferroelectric and offers higher capacitance ranges than Class I.

Y5V (Z5U)

The Y5V is a general purpose Class II dielectric with a higher dielectric constant and a greater variation of properties with temperature and test condition. Very high capacitance per unit volume is attainable for general-purpose applications.

Capacitors are supplied in tape-and-reel or magazine packaging typical of resistors. When specifying capacitors for reflow- or wave-solder processing, it is important to select the proper end-cap termination material. The most compatible end-cap termination plating is the nickel barrier/solder type.

Standard molded tantalum capacitors

Manufacturers have defined standards for a plastic-molded surface mount tantalum capacitor. Other sizes and shapes are on the market, each with excellent mechanical and electrical characteristics, but only the uniform-molded body type shown in Fig. 2-8 meets the criteria as an industry-recognized standard.

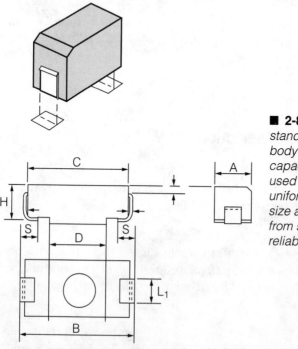

■ **2-8** *The EIA standard molded body tantalum capacitor is widely used because of its uniform physical size and availability from several reliable sources.*

The basis for these standards was influenced by increased worldwide demand and the need for multiple sources by the users. The EIA standard-molded tantalum capacitor shown is preferred by assembly specialists because of the uniform package dimensions and flat surfaces. The flat surface is ideal for automated assembly systems using vacuum for pickup and placement. The physical dimensions of the four basic package sizes are associated with the dielectric mass and operating voltage (see Table 2-2).

■ Table 2-2 Physical factors controlling the EIA standard molded tantalum capacitor family.

Case code	Length	Body width	Height	Glue pad	Termination Length	Width	Height
3216	3.2±0.2	1.6±0.2	1.8±0.2	0.8 Min	0.8±0.3	1.2±0.1	0.7 Min
3528	3.5±0.2	2.8±0.2	1.9±0.2	1.1 Min	0.8±0.3	2.2±0.1	0.7 Min
6032	6.0±0.3	3.2±0.3	2.5±0.3	2.5 Min	1.3±0.3	2.2±0.1	1.0 Min
7343	7.3±0.3	4.3±0.3	2.8±0.3	3.8 Min	1.3±0.3	2.4±0.1	1.0 Min

Levels of electrical and mechanical limits are required with each configuration, and because of their uniform size they must be compatible with the automated assembly systems.

The standards established by the EIA for tantalum capacitors include two variations of component specifications. The standard capacitance range, shown in Table 2-3, includes values from 0.10 uF through 68.0 µF. The extended range, shown in Table 2-4 provides capacitance values through 330 µF.

■ Table 2-3 Standard value range and operating voltage for the EIA molded tantalum device family.

	Max capacitance (µF)/voltage/case size							
	voltage							
Case size	4	6	10	16	20	25	35	50
3216	3.3	2.2	1.5	1.0	.68	.47	.33	.10
3528	10.0	6.8	4.7	3.3	2.2	1.5	1.0	.33
6032	33.0	15.0	10.0	6.8	4.7	3.3	2.2	1.0
7343	68.0	47.0	33.0	22.0	15.0	10.0	4.7	2.2

■ Table 2-4 The extended range device allows for higher capacitance value and operating voltage.

| Case size | \
Max capacitance (μF)/voltage/case size \
voltage | | | | | | | |
	4	6	10	16	20	25	35	50
3216	10.0	6.8	4.7	3.3	2.2	1.5	1.0	.33
3528	15.0	15.0	6.8	4.7	3.3	2.2	1.5	.68
6032	100.0	68.0	47.0	33.0	22.0	15.0	10.0	3.3
7343	330.0	150.0	68.0	47.0	47.0	33.0	22.0	6.8

Nonstandard miniature tantalum devices

Nonstandard device types are usually available from a single source. Miniature tantalum capacitors, for example, are available for high-density applications, such as multichip modules, but secondary suppliers of these devices might not be available.

The miniature tantalum device is not as uniform in physical size as the molded types shown above, but high-density and multichip module applications are usually restricted in area and overall height and need the smallest component size available. Because the dielectric volume is different for each value, the epoxy-coated tantalum device can be dimensionally different from one another and the end cap termination or contact is not always uniform in size. Process engineers have noted that the small wire protruding from the positive end of the epoxy-coated tantalum device occasionally hangs up in tape-and-reel packaging systems.

Inductor devices for SMT

Standards for surface mount inductors are not as defined as with most of the passive devices. These devices can have unique physical characteristics, depending on the manufacturer, the materials, and the manufacturing techniques.

Figure 2-9 details three inductor package types. Although all configurations available from the industry are not shown here, these are typical for the chip, precision wire wound, and molded products on the market.

Standard surface mount diodes

Diodes are furnished in a variety of package types. All of those shown in Fig. 2-10 are standard. Depending on the specific device type and use category, the physical shapes might be quite different from each other. Because of the similarity between the basic

26

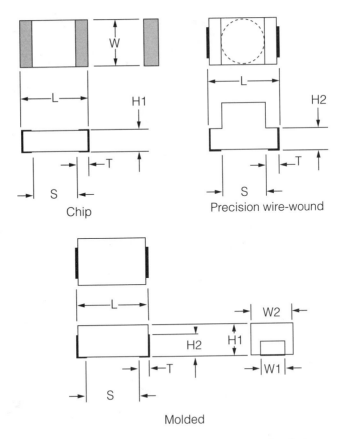

Chip

Precision wire-wound

Molded

■ **2-9** *Inductor families are not controlled by industry standards and value-to-size or shape will vary from one manufacturer to another.*

"die" of these devices, many diodes share the same package type as SMT transistors.

A variation of the SOT-143 for device packaging is available from manufacturers in Japan. The user should note that the Japan-manufactured device is wider overall than the United States-supplied device and the wide lead that identifies pin 1 is moved to the lower end of the package.

Standard transistor packaging for SMT

Transistor standards are well established in the industry and the designer has access to scores of manufacturers willing to supply

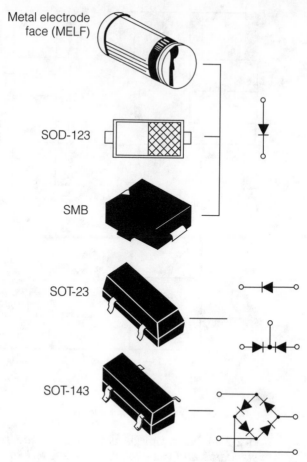

Metal electrode
face (MELF)

SOD-123

SMB

SOT-23

SOT-143

■ **2-10** *Diode packaging standards allow for diverse operating conditions and unique physical configurations.*

general-purpose devices in common package styles. The SOT-23 is the most widely used package; size and lead style of the transistor will be different for higher power applications requiring enhanced thermal management. In addition to the examples of transistor packaging shown in Fig. 2-11, transistor array (multi-unit) packaging is also available in surface mount.

The most common case types used for Bipolar and Field Effect Transistors (FETs) are SOT-23, SOT-24, and SOT-89. The SOT-23 and SOT-24 are smaller, with three or four leads respectively.

The designer converting from leaded technology to SMT should be aware that the pin assignments of the SOT transistor contacts are not the same as the TO-5 or TO-92 (lead-type) devices. In addi-

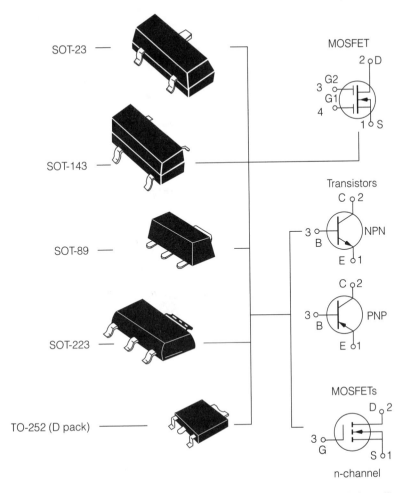

SOT-23

SOT-143

SOT-89

SOT-223

TO-252 (D pack)

MOSFET

Transistors

NPN

PNP

MOSFETs

n-channel

■ 2-11 *Transistor packaging for surface mount is designed for efficient power management as well as compatibility with most SMT assembly processes.*

tion, the internal wire bonding might vary with some manufacturers. Study the pin assignment specifications carefully before designing the PC board. FET pin assignment can vary from one device type to another, even with the same manufacturer.

Transistor array packaging

Transistor and diode arrays are offered in SOIC- and PLCC-type housings, but multiple sources are not always interchangeable. The MELF component style, which is available in the standard tape-and-reel packaging, is often used for single diodes and rectifiers. Size will vary with AMP rating, type, and manufacturer. See Fig. 2-12.

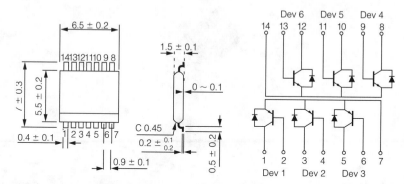

■ **2-12** *A typical surface mount transistor array package.*

A miniature transistor array package is available from some off-shore suppliers and might offer a solution for high-density PC board applications. The smaller SO package will accommodate several configurations of low-power, general-purpose transistors in a gull-wing lead plastic body about ¼ the size of the conventional SOIC package.

Integrated circuits for SMT

Devices classified as active are the integrated circuits (ICs), which require several terminations for attachment to a PC board. A wide variety of package families exist for the IC and all are considered standard, but because of specific applications and lead count they are physically different.

Although standards have been well established for most surface mount devices, component engineers, designers, and purchasing personnel must work together to ensure that the correct component is designed into and purchased for the product. The success of an SMT project is directly linked to the quality of communication between all members of the development team.

The more common IC package type used for high-volume applications typical of logic, analog, timers, and driver circuits is the Small Outline or SOIC with a formed Gull-Wing lead design. Memory, Programmable Logic, and Gate Array ICs are often furnished in a plastic-leaded chip carrier (PLCC) and SOJ package having a re-

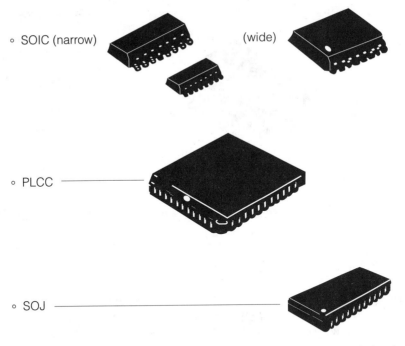

SOIC (narrow)   (wide)

PLCC

SOJ

■ **2-13** *Surface mount IC packages are furnished with two basic lead shape styles, gull wing lead typical of that shown on the SOIC and J formed lead typical of PLCC and SOJ devices.*

verse "J" lead for termination. Figure 2-13 illustrates the variety of shapes and lead designs developed for IC packaging.

IC packaging standards

Commercial surface mount ICs are available in several standard body and lead frame styles: Small Outline Gull Wing (SO); Small Outline J-lead (SOJ); Plastic Lead Chip Carrier (PLCC); and Quad Flat Pack (QFP). While all manufacturers do not conform to a single configuration, standards have been set for most of the IC families. The EIA/JEDEC registered IC component types are available in both inch and metric sizes.

Small outline IC

The SOIC device shown in Fig. 2-14 is offered by various United States and off-shore IC manufacturers. Most of the domestic manufacturers' configurations meet the JEDEC registered specification. The dimensions shown reflect the JEDEC standard.

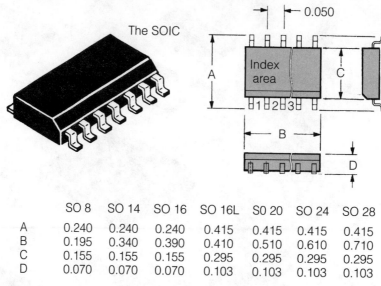

The SOIC

0.050

Index area

A

B

C

D

1 2 3

	SO 8	SO 14	SO 16	SO 16L	SO 20	SO 24	SO 28
A	0.240	0.240	0.240	0.415	0.415	0.415	0.415
B	0.195	0.340	0.390	0.410	0.510	0.610	0.710
C	0.155	0.155	0.155	0.295	0.295	0.295	0.295
D	0.070	0.070	0.070	0.103	0.103	0.103	0.103

Dimensions shown are nominal

■ **2-14** *The JEDEC registered small outline IC package has a gull-wing lead shape on 0.050-in center-to-center spacing and is standard.*

The JEDEC-registered Small Outline, or SOIC contact pattern, is easily adaptable to the circuit board. The two parallel rows of contacts have the same pin assignment as the dual in-line through-hole IC they replace. Most logic devices, for example, maintain a consistent spacing between contact row centers. However, depending on the manufacturer, the same IC function can be housed in an entirely different component style for SMT. The designer needs to review specifications before routing signal traces as the pinouts will vary significantly from one package style to another.

An example of the same device function housed in the PLCC 20 and SO-14 is shown in Fig. 2-15. In comparing these device types, the designer should note the SOIC has the same pinout as the conventional DIP, while the PLCC-20 has a unique pinout with non-connected pins dispersed on each of the four sides.

Some SO devices require a wider lead frame to provide for the die size or the heat dissipation of the silicon. A PC board might have a mix of the SO-16 and SO-16L (wide) devices. Always check the component manufacturer's specifications. SOICs greater than 16 leads will be in the wider lead frame.

In Japan, the counterpart to the United States JEDEC committee is the EIAJ. While manufacturers outside the United States use the metric system, they do maintain the 0.050-inch lead spacing

Typical 74LSO2 device

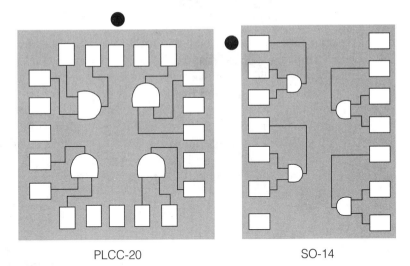

PLCC-20   SO-14

■ **2-15** *The same device functions might be available in optional package styles. The component shape and pin assignment will be very different from one device type to another.*

Leads	SO-N	SO-W	SO-X	SOP
6				6.99
8	6.20	10.60		6.99
14	6.20	10.60		6.99
16	6.20	10.60	12.16	8.89
18			12.16	8.89
20		10.60		8.89
24		10.60	12.16	10.80
28		10.60	12.16	12.70
30				12.70
32		10.64		14.61
36		10.60		14.61
40				16.51
42				16.51

■ **2-16** *Dimensions shown compare the maximum toe-to-toe width of the JEDEC standard SOIC outlines to the EIAJ standard SOP device outline.*

on many IC components to be compatible with the established American market. Standard SOIC packages are for both JEDEC and EIAJ and are shown in Fig. 2-16.

Significant mechanical differences remain on Japanese products available with 1.27 mm (0.050 in) lead spacing. As an example, the distance between rows on the contacts of the EIAJ-SOP is not the same as the JEDEC SOIC. When specifying IC devices from multi-

Component selection for SMT

ple sources, it might not be possible to mount both offshore and domestic sizes on the land pattern or pad geometry recommended by only one component manufacturer.

Manufacturers of the Small Outline Plastic (SOP) IC in Japan do vary from the JEDEC standard (Table 2-4), and even though the 1.27-mm (0.050-in) contact spacing is the same as the JEDEC part, the body width and distance between contact rows is often inconsistent from one source to another. If the circuit board is designed to meet JEDEC SOIC specifications, and an EIAJ SOP part is substituted on the assembly, the leads of the device will overlap the narrower JEDEC land pattern. A land pattern design that might provide a simple solution for this multiwidth situation is illustrated in chapter 3.

SOIC

The gull-wing form of the SOIC device provides an overall low profile, allowing for ease of visual solder joint inspection. With the correct land pattern geometry, the device will self-align during the reflow soldering process.

Thin small outline package (TSOP)

Like the SOIC packages in wide use today, the TSOP has two rows of gull-wing formed leads on opposing sides. Except for the gull-wing lead shape, TSOP and SOIC packages are physically different.

Leads of the SOIC are on the long edge of the rectangular device with 0.050-in pitch. The TSOP family is designed for lead exit at the end of the rectangular shape as shown in Fig. 2-17. Four package sizes are established as standard for the TSOP configuration to accommodate a variety of die size and I/O. Table 2-5 describes the nominal dimensions for the TSOP.

■ Table 2-5 Comparing TSOP body
size to lead pitch and typical I/O for deivce.

Body size W × L	@ .65 mm (0.0256 in.)	@ 0.5 mm (0.0197 in.)	@ 0.4 mm (0.0158 in.)	@ 0.3 mm (0.0118 in.)
6.0 × 12.4 mm	16	24	28	36
8.0 × 14.4 mm	24	32	40	52
10.0 × 16.4 mm	28	40	48	64
12.0 × 18.4 mm	36	48	60	76

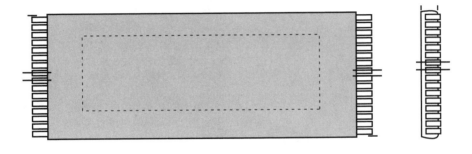

■ **2-17** *The contact leads of the TSOP are located at the device ends to accommodate the rectangular chip design typical of memory devices.*

Developed to meet the low-profile needs of the PCMCIA and products having a restricted overall height, the TSOP is widely used
throughout the industry. The overall package height from the
mounting surface is 1.27 mm (0.050 in), which is far less than the
3.5 mm (0.138 in) of the SOJ device. Having all leads at opposite
ends of the package accommodates routing of the circuit and has
proven ideal as an option to the SOJ package for memory products. Several manufacturers have adapted the TSOP for a series of
1 MEG and 4 MEG Flash Memory products.

The smart card standards limit the overall thickness of the PCM-
CIA assembly. The thin devices, generally attached to a 0.5-mm
(0.020-in) thick substrate, can stay within the limits established
and even allow TSOP attachment on both primary and secondary
sides of the substrate. The PCMCIA meets the physical standard
established by the Personal Computer Memory Card Industry Association.

Plastic lead chip carriers and quad packages (PLCC)

Several standard J-lead packages are registered with JEDEC and
are available from many domestic, European, and Asian component manufacturers. Devices furnished with the "J" lead form
(PLCC and SOJ) are higher in profile furnishing a greater standoff
height from the board surface. The PLCC and SOJ standoff height

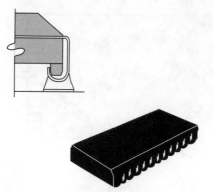

■ **2-18** *The SOJ IC widely used for memory applications is available in four standard widths: 0.300 in, 0.300 in, 0.400 in, and 0.450 in.*

offers a better clearance for cleaning but the "J" design is susceptible to damage causing problems in maintaining coplanarity (see the details in Fig. 2-18).

The square plastic lead chip carrier (PLCC) has an equal number of contacts on each of its four sides. The contact is formed in a J shape extending from the body, formed down and under the package. This lead shape offers a reliable attachment while decreasing the surface area required for each device.

The uniform sizes and J-lead design of the PLCC have been well received by the users of custom and semicustom ASIC devices and the clearance under the PLCC allows for efficient cleaning after solder reflow.

Package sizes greater than 84 leads are not recommended because they are too difficult to place using robotic assembly. The assembly machines are accurate enough, but the larger parts are bulky in size and maintaining lead coplanarity is difficult.

Manufacturers of the PLCC also complain of manufacturing yield problems on the high (100 and 124) lead-count devices due to excessive internal wire bond length. The delicate wire must span a greater distance from the silicon die to the lead frame, and during the plastic molding operation, wire bonds are often washed away by the pressure of the molding process.

Another style of package similar to the PLCC is the "C" pack. The leads are on 1.27 mm (0.050 in) or less spacing on four sides, but instead of the J-bend, the leads are bent straight down to the board surface. This "butt joint" style is not standard and requires a unique land pattern for mounting. See Fig. 2-19.

Some PLCC products are furnished in a rectangular package as well. The PLCC-18 for example, is supplied in two sizes and each

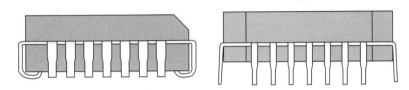

■ 2-19 *The J-lead PLCC is considered a standard for the four-sided ICs up to 84 pins. Some manufacturers furnish a similar but nonstandard device in the I-lead or butt-mounted configuration.*

has a rectangular shape with four leads on each end and five leads per side. The width of the IC is consistent, but the length will vary more than 1.5 mm (0.060 in). Details are shown in Fig. 2-20.

SOJ (small outline J-lead) package

Originally developed to package memory devices (DRAM), the SOJ configuration has proven to be a popular alternative for other device packaging requirements as well. Unlike most of the newer generations of surface mount devices, the SOJ was developed using "inch" measurements and is generally referred to by the package width (i.e., 0.300 in, 0.400 in). The body size is comparable to the DIP IC except that the lead pitch is 1.27 mm (0.050 in). When compared to the SOIC package the designer should notice that although the lead pitch is the same, the overall SOJ package height is one and a half times greater and the overall surface area required to attach a 28 lead (0.300-in wide) device is significantly less.

The standoff height or distance between the board surface and the bottom surface of the SOJ lead devices is typically 0.9 to 1.1 mm (0.035 to 0.045 in). Considering the body mass and standoff height, the SOJ device will have excellent thermal characteristics. For those applications requiring very close side-by-side spacing typical of the single in-line memory module, a special bypass capacitor family has been developed for solder attachment under the SOJ device. The SOJ IC package is available in four standard sizes defined by inch measures, 0.300-in, 0.350-in, 0.400-in, and 0.500-in wide.

Packages in wider sizes can also be adapted to accommodate the higher performance products needing space for larger die sizes. One note of caution, data sheets furnished by component suppliers might only use a single physical example of the SOJ with a subtle reference to package width buried in the text or in tables not associated with the mechanical detail.

The PLCC
No. of leads

	A	B
20	0.395	0.330
28	0.495	0.430
44	0.695	0.630
52	0.795	0.730
68	0.995	0.930
84	1.995	1.130
100	2.195	1.330
124	2.495	1.630

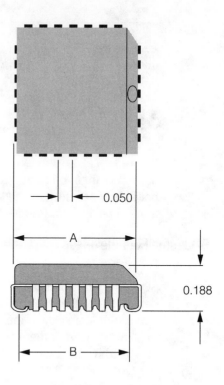

0.050

A

0.188

B

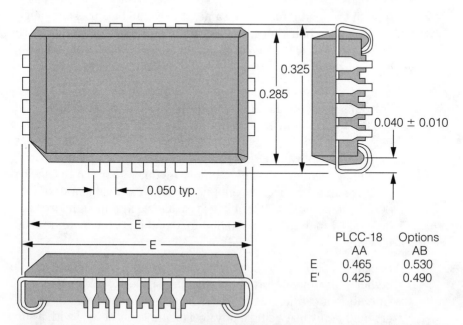

0.325

0.285

0.040 ± 0.010

0.050 typ.

E

E

	PLCC-18	Options
	AA	AB
E	0.465	0.530
E'	0.425	0.490

■ **2-20** *The JEDEC-registered PLCC IC package has 0.50-in center-to-center lead spacing on four sides. This package is a common choice for custom and semicustom gate-array logic devices.*

Quad flat pack for SMT

Due to technological advances and the growing complexity of silicon products it is often necessary to furnish more than one function within a single package. Current generations of high-performance electronic devices require increasing lead counts or I/O and closer lead spacing (pitch). In doing so the overall size of the component must be kept to a practical limit.

The industry standards for fine-pitch components and component packaging might seem fragmented and even the land print geometry recommended by manufacturers appear to be in conflict. However, standards are in place and the QFP family has expanded to meet specific applications and physical restrictions. Industry standards are in place for a wide selection of plastic Quad Flat Pack devices. These standards, developed somewhat independently by JEDEC and EIAJ members, are now recognized universally by IC manufacturers worldwide. In the following text, component families with 0.025 in (0.63 mm), 0.5-mm (0.020 in), 0.4-mm (0.016-in), and 0.3-mm (0.012-in) lead pitch will be defined.

JEDEC 0.025-in lead pitch QFP

The category of fine pitch refers to four sided, gull-wing lead device package with 0.025 in (0.63 mm) center-to-center lead spacing or less. A typical device defined as a JEDEC Standard QFP is shown in Fig. 2-21. The Quad Flat Pack(age) developed using "inch" measurements is available in both square or rectangular configurations.

The device standard defines a symmetrical 0.025-in lead pitch and offers the manufacturer of the devices the option of corner extensions or a barrier for physically protecting the leads from damage during handling or assembly processing. Even though the leads are physically protected, bent leads on fine-pitch devices will continue to be a problem. In addition, because the corner barriers are optional, most IC manufacturers rely on a partitioned carrier tray for QFP handling with features built into the carrier to protect the leads.

Standard quad flat pack devices are also available with lead pitch both greater and less than 0.025 in. For example, QFP devices with the 1.0-mm (0.040-in) and 0.8-mm (0.031-in) lead spacing is available for custom and semicustom IC applications but newer, high-performance devices are more likely furnished in a lead pitch of 0.5 mm (0.2 in) or less. These families are developed using the

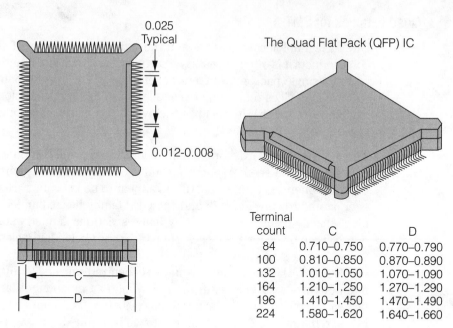

0.025
Typical

0.012–0.008

The Quad Flat Pack (QFP) IC

Terminal count	C	D
84	0.710–0.750	0.770–0.790
100	0.810–0.850	0.870–0.890
132	1.010–1.050	1.070–1.090
164	1.210–1.250	1.270–1.290
196	1.410–1.450	1.470–1.490
224	1.580–1.620	1.640–1.660

■ **2-21** *JEDEC-registered standard for the quad flat pack IC defines a gull-wing lead pattern with 0.025-in center-to-center lead spacing.*

40

international "metric" (millimeter) rather than in inches, providing a more precise physical tolerance for the smaller features.

EIAJ/JEDEC Shrink Quad Flat Pack (SQFP)

One of the fastest growing families of fine-pitch products is a package identified as SQFP or Shrink Quad Flat Pack. Both a square and a rectangular package are offered for the SQFP and the standard pitch options includes 0.5 mm, 0.4 mm, and 0.3 mm. The 0.3 mm is being referred to in some circles as ultra-fine pitch, but the term "ultra" might be ignored as they become more common.

Square SQFP device

Typical of the QFP, the gull-wing leads extend away from each of the four edges. As noted, three lead pitch options are offered and the package size range for the square SQFP is from 5.0 mm (0.197 in) through 44.0 mm (1.74 in).

As an example, Fig. 2-22, illustrates one of the package configurations available. The maximum I/O for the 44.0-mm (1.74-in) square package with 0.3-mm (0.012-in) pitch is 576 total leads. It is expected that with all the choices available in the standard, this

■ **2-22** *The EIAJ standard includes both rectangular and square package shapes and optional lead pitch.*

low-profile package family will accommodate the silicon chips for current usage, as well as the foreseeable future. See Table 2-6.

■ **Table 2-6 Compares body size and lead pitch to I/O capacity for the square SQFP device family.**

Body size W × L	@ 0.5 mm (0.019 in.)	@ 0.4 mm (0.015 in.)	@ 0.3 mm (0.012 in.)
5.0 mm	24/32	32/40	48/56
6.0 mm	32/40	40/48	56/64
7.0 mm	40/48	56/64	72/80
10.0 mm	64/72	80/88	112/120
12.0 mm	80/88	100/108	136/144
14.0 mm	100/108	120/128	168/176
20.0 mm	144/152	176/192	248/256
24.0 mm	176/184	216/232	296/304
28.0 mm	208/216	256/272	352/360
32.0 mm	240 /248	296/312	400/408
36.0 mm	272/280	336/352	456/464
40.0 mm	304/312	376/392	512/520
44.0 mm	336/344	424/432	568/576

Rectangular SQFP devices

Six rectangular packages are defined in the EIAJ/JEDEC standard. The rectangular SQFP has the same lead space options as the square version, but the lead counts reflect the uneven side dimension. Rectangular lead frame design is offered for packaging the nonsymmetrical silicon die shapes. Table 2-7 reflects the rectangular size and pitch ratio to total lead count.

■ Table 2-7 Compares body size and lead pitch to I/O capacity for the rectangular shape SQFP.

Body size W × L	@ 0.5 mm (0.019 in.)	@ 0.4 mm (0.015 in.)	@ 0.3 mm (0.012 in.)
5.0 mm × 7.0 mm	40	52	68
7.0 mm × 10.0 mm	60	76	100
10.0 mm × 14.0 mm	88	108	148
14.0 mm × 20.0 mm	128	160	216
20.0 mm × 28.0 mm	184	232	308
28.0 mm × 40.0 mm	264	332	440

While the use of miniature fine-pitch devices has provided layout flexibility, the need to push components even closer together on very complex multilayer substrates has also impacted the cost of the product.

Companies adapting fine pitch are attempting to relieve the growing space dilemma by combining the function of several integrated circuits into one or two multi-function custom devices.

The customized or application specific ICs might alleviate the PC board grid lock, but higher lead count and finer lead pitch will generally compel the designer to use more circuit layers on the PC board, thus increasing PC board fabrication cost.

Ball Grid Array (BGA) packaging

The Ball Grid Array concept for high I/O devices is gaining a great deal of momentum. Advanced ICs require improved performance, better thermal management, and higher circuit density. The component industry has adapted a family of BGA devices with the potential for more than 1,000 I/O contacts. The routing of circuit traces of the higher pin count devices will be a challenge but the 200–300 contact range will prove both feasible and manageable. The contacts are not leads or pins but small 0.8 mm (0.031 in) diameter tin/lead alloy spheres attached to the bottom of the device.

The BGA package style is ideal for attachment of high-performance ICs and a reliable alternative to lead frame termination. In addition, because of the rugged construction of ball grid packaging, handling will be less stringent than the fine-pitch devices.

The advantages of Ball Grid Array packaging include:

☐ High-density interconnection capability
☐ Rugged construction

☐ Electrical performance

☐ Excellent thermal management

☐ Fast design to production schedule

☐ High reliability

☐ Compatible with current reflow soldering

☐ Devices can be pretested

☐ Improved performance with shorter circuit paths

Concerns of Ball Grid Array technology include:

☐ Impossible to visually inspect solder joint

☐ Rework and repair equipment is specialized

☐ Larger devices might not remain flat

☐ Reattachment of BGA device might be difficult

One of the most attractive elements of the BGA concept is for replacing fine-pitch leaded devices and the elimination of bent or deformed leads. Physical damage to leads is the most common cause of solder defects on fine-pitch devices. It only takes one lead formed too far up or too far down to cause a major solder joint defect. If the BGA can remove that defect element, then Six Sigma yields might be a reality for everyone. The packaging system has been used successfully in high-end computer products for several years. The original products were developed using co-fired ceramic as a base material but the commercial products are furnished today using plastics.

Standards for BGA packaging

Physical standards for Ball Grid Array packaging is established by the industry members of JEDEC (JC-11). Although custom MCM applications are not necessarily restricted to meet the registered stand package, designers will find the specifications adaptable to almost any application. The JEDEC standards have included both plastic and ceramic package configurations to meet commercial as well as hi-rel applications.

Plastic Ball Grid Array

The plastic BGA configuration uses a laminate-base material typical of that used to manufacture conventional PC boards. An ideal mounting surface for attaching the IC die, the concept has proved to be a reliable, cost effective, and space efficient alternative to etched lead frame interface technology. Assembly process specialists are very satisfied because the parts will actually self-align to the land patterns during reflow assembly processing.

Standard contact matrix

The lead pitch and contact count are defined by using a matrix guideline established in the standard. Physical size of the package will be determined by the number of I/O required, lead or contact pitch, and array pattern. The standard allows for both full array (uniform rows and columns) and a staggered array pattern. The standard pitch options include 1.00 mm, (0.040 in), 1.27 mm (0.050 in), and 1.50 mm (0.060 in). See Fig. 2-23.

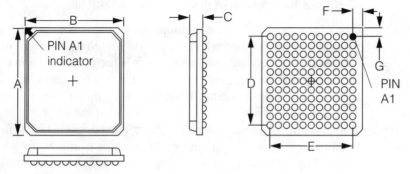

■ **2-23** *Ball Grid Array device offers a great deal of latitude in I/O capacity in a relatively rugged and physically small package outline.*

When planning the I/O requirement for the BGA device, consider the standard configurations and numbering system established in JEDEC.

JEDEC (JC-11) BGA standards include:

☐ EIA/JEDEC MO-151 PBGA Plastic Ball Grid Array

☐ EIA/JEDEC MO-156 CBGA Ceramic Ball Grid Array

☐ EIA/JEDEC MO-158 CCGA Ceramic Column Grid Array

☐ EIA/JEDEC MO-163 R-PBGA-B Rectangular Plastic BGA

Ceramic Column Grid package

The devices defined as Column Grid Array are generally the same as the BGA except a round post contact is furnished for attachment in place of the sphere contact. The post configuration has been provided to allow attachment of the higher performance ceramic package to conventional laminate materials. The post or column lead adds a level of compliance during thermal cycling reducing physical stress to the solder joint.

Chip Scale(CSP) packaging

In the DGA process, the IC die is mounted face down to a complaint interface specifically designed for mating with the die attachment sites. The intermediate structure also adapts polyimide films as a base carrier for the conductor circuit. As in BGA technology, the DGA package is assembled and tested as a component before attachment to the board. Using an additive "pure gold" alloy process, the conductors are routed inward or under the silicon die to a uniform grid pattern. The spacing of the grid might vary to meet the I/O requirements of each device but 0.5-mm (0.020-in) pitch is common. The grid pattern is furnished with either a bump or ball of nickel/gold alloy that is compatible with both solder and conductive polymer attachment. See Fig. 2-24.

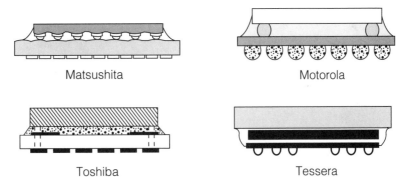

Matsushita　　　　　Motorola

Toshiba　　　　　Tessera

■ **2-24** *Chip Scale Packiaging (CSP) typical of the examples shown can provide a finished device outline that is the same size or near the size of the die itself.*

Land pattern development for SMT

Land pattern planning

THE LAND PATTERN GEOMETRY AND SPACING FOR
attaching surface mount devices varies due to package shape, lead
spacing, and contact type. In this chapter, proven land patterns for
each of the component types presently used are illustrated and
guidelines are presented for creating suitable patterns for future
products.

Land pattern geometry and spacing follow the recommendations
and requirements furnished by IPC, EIA, and leading component
manufacturers. Geometries presented here should provide the
process engineer with an opportunity to yield a robust solder joint.
The designer and process engineer might agree to make adjustments to those land patterns when it is warranted. Subtle changes
in contact shapes will evolve due to improvement in assembly
processes or refined component quality. The contact geometry
shown in this chapter is suited for the reflow-solder process, but
variations are also illustrated for wave-solder applications.

While component manufacturers will often furnish geometry and
spacing for the contact land pattern, they are not always familiar
with the CAD technology used to develop photo tooling and PC
board fabrication. When component manufacturers prepare the
land pattern recommendation, they might not be aware of the sig-
nificance of grid location for device placement and circuit routing.

When developing the CAD database, the designer might tailor the
manufacturer's recommended geometry by adjusting for grid positioning of the contact land patterns. These refinements will assist
the designer in both component placement and trace routing.

Figure 3-1 describes the typical land pattern for the JEDEC-registered plastic leaded chip carrier. The 0.63-mm (0.025-in) contact

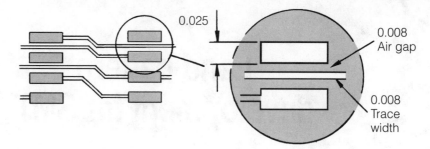

0.025

0.008
Air gap

0.008
Trace
width

■ **3-1** *Conductor routing between 0.050-in spaced land patterns is common. The land pattern geometry to conductor trace spacing should allow for a minimum air gap of 0.008 in.*

width will allow for the routing of 0.20-mm (0.008-in) wide conductor traces between leads if needed.

If length is added to the contact pattern, it might provide an even more robust solder joint, but it might be necessary to reposition or adjust the grid center again.

Passive component land pattern design

While resistors and capacitors are available in a wide choice of sizes, the designer should attempt to select a common size for general use. A uniform component type and land pattern gives the assembly specialist greater control over the equipment and processes used. Although miniature products require the use of small 0402 or 0603 devices to achieve the design goals, larger devices will provide a wider value range, especially for capacitors. Most low- to mid-range capacitors are available from several sources in the 0805, 1206, and 1210 configurations.

As noted in chapter 2, the industry has standardized the chip component family and recommends seven resistor sizes and eight capacitor shapes. Assembly specialists prefer 0805 or 1206 sizes for resistors and the 0805-, 1206-, or 1210-size capacitor because these resistor and capacitor sizes are easier to handle than the 0603 or smaller devices.

Due to the variables of value, dielectric, and power applications, several shapes are available. To meet the higher volume commercial needs of the industry and sustain a reasonable supply in distribution, reducing the number of size options was necessary. Figure 3-2 reflects the part size configuration and footprint geometry recommended for each of the standard devices. As with the

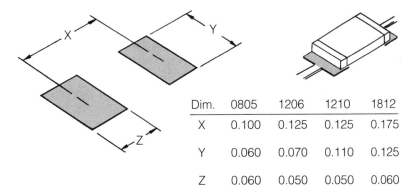

Dim.	0805	1206	1210	1812
X	0.100	0.125	0.125	0.175
Y	0.060	0.070	0.110	0.125
Z	0.060	0.050	0.050	0.060

■ **3-2** *Center line location for SMT land pattern features will assist the designer in database construction of the CAD library.*

resistor family, the dimensional limits shown allow for the manufacturing variables of different suppliers.

The dielectric types of COG, X7R, and Z5U are the most common per EIA Standard RS-19 and are available in various sizes considered standard 1608 (0603), 2012 (0805), 3216 (1206), 3225 (1210), and 4532 (1812). All sizes shown are considered acceptable for efficient assembly processing but due to excessive cracking, the 4564 (1825) device should be avoided.

The selection of value and size should be made early in the design because land patterns and geometry of each device type is very different. See Fig. 3-3 for details.

When developing the land pattern the designer must consider both tolerance limits of the component and assembly process characteristics. Width of the land pattern for chip components is the first consideration for reflow-solder processing. If the pattern is too wide, the chip component could rotate during the solder process as illustrated in Fig. 3-4.

Likewise, a land pattern that is too long will cause the chip component to float or pull away from the land pattern during cooling of the liquid solder material. The ideal pattern allows the liquid solder to heat and cool evenly providing equal surface tension. When surface tension is equal, the component will center on both contact land patterns.

The land pattern shown in Fig. 3-5 provides the most satisfactory results in the reflow process and meets the requirements of the J-STD-001 solder attachment standard.

49

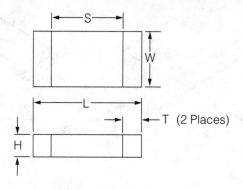

	L	W	T	H
0805/R	0.079 0.073/0.085	0.049 0.043/0.055	0.016 0.006/0.026	0.020 0.014/0.026
0805/C	0.079 0.071/0.087	0.049 0.041/0.057	0.020 0.010/0.030	0.035 0.030/0.043
1206/R	0.126 0.120/0.132	0.063 0.057/0.069	0.020 0.010/0.030	0.022 0.016/0.028
1206/C	0.126 0.118/0.134	0.063 0.055/0.071	0.020 0.010/0.030	0.043 0.039/0.053
1210/R	0.126 0.120/0.132	0.098 0.093/0.104	0.020 0.010/0.030	0.022 0.016/0.028
1210/C	0.126 0.118/0.134	0.098 0.091/0.106	0.020 0.010/0.030	0.043 0.039/0.053

■ **3-3** *Physical limits for 0805, 1206, and 1210 chip devices.*

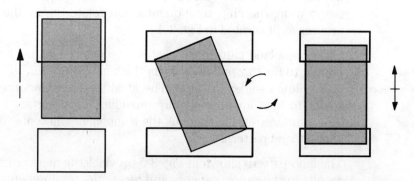

■ **3-4** *The land pattern geometry is specifically designed to promote self-centering of each component while the assembly is exposed to the reflow-solder process.*

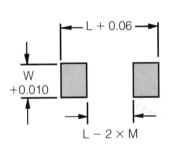
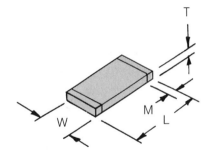

■ 3-5 *Land pattern formula for chip resistors or capacitors has been developed for a stress resistant, reliable, and uniform solder termination.*

The land pattern geometry should promote self-alignment during reflow, even when the placement of the component is not accurate. Ideally, a finished solder connection extends or wicks onto the end-cap area by 25–50% of the overall component height, as shown in Fig. 3-6.

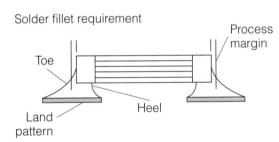

■ 3-6 *Land pattern must be calculated to meet minimum solder joint profile requirements and allow for both component and assembly process tolerances.*

As mentioned previously, resistors and capacitors are available in many sizes. Land patterns illustrated here will meet process requirements for each of the sizes recommended as standard. Although the contact design and spacing is typical of the formula shown in Fig. 3-7, this basic formula is adaptable to any chip style device.

Optional wave-solder land pattern design for passive components

Attachment of chip components and wave soldering on the secondary side of the PC board is common, especially for PC boards with mixed technology. Often a company's first application of SMT employs this technique because placement equipment cost is not prohibitive and wave-solder machines might be in use already.

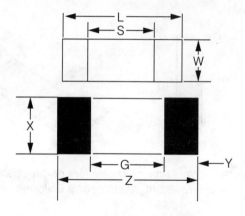

3-7 *A typical basic formula for many chip style devices.*

C = Component geometric tolerance

F = Land pattern geometric tolerance

P = Placement equipment accuracy (DTP)

J = 2 Times solder joint requirement for toe or heel

$$Z = L + J + \sqrt{C^2 + F^2 + P^2}$$

$$G = S - J - \sqrt{C^2 + F^2 + P^2}$$

The contact pattern geometry developed for reflow solder processing noted above can work well in wave solder; however, some companies might choose a narrow contact area to restrict excessive solder buildup.

For those using wave solder to terminate chip components, the devices are attached first with epoxy, and because solder will be used primarily for electrical connections, it is acceptable to reduce the width of the contact pad geometry. Although the pattern shown previously works well in the wave-solder of the very small 0402 and 0603 devices, the narrow (by 30%) pattern, illustrated in Fig. 3-8, minimizes excess solder on 0805 and larger devices.

When the designer is laying out the PC board, the narrow land pattern geometry appears to allow generous spacing. However, clearances between component bodies must be maintained. In addition, when component bodies are too close to an adjacent contact lead or exposed via pad, bridging can occur during wave solder as shown in Fig. 3-9.

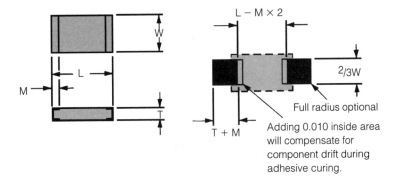

L − M × 2

W

L

M

T

2/3W

Full radius optional

T + M

Adding 0.010 inside area
will compensate for
component drift during
adhesive curing.

■ **3-8** *Excessive solder buildup of epoxy-attached devices can be
avoided during wave solder by reducing the width of the land pattern.*

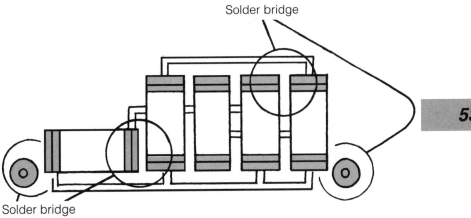

Solder bridge

Solder bridge

■ **3-9** *Spacing between chip components must be adequate to avoid
bridging during wave-solder processing.*

The contact geometry, detailed in Fig. 3-10, is specifically designed for epoxy-attached components processed in a wave-solder
system.

Compensate for the component body as well as the contact area
when using the narrow land pattern. When spacing the device, allow an additional 0.8- to 1.0-mm (0.03- to 0.04-in) clearance to an
adjacent component or via hole pad. Wave soldering of large 2225
chip components is not advisable due to the high occurrence of
cracking caused by thermal mismatch and flexing of the substrate.

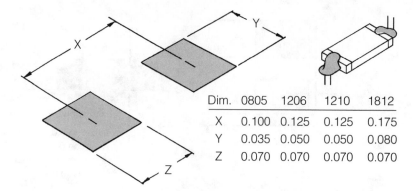

Dim.	0805	1206	1210	1812
X	0.100	0.125	0.125	0.175
Y	0.035	0.050	0.050	0.080
Z	0.070	0.070	0.070	0.070

■ **3-10** *The optional narrow land pattern for chip resistors and capacitors should be used only for wave solder of epoxy-attached components.*

MELF component land pattern geometry

The MELF (Metal Electrode Lead Face) is a round-shaped component dimensionally similar to the chip devices and will adapt to the standard patterns detailed in Fig. 3-11.

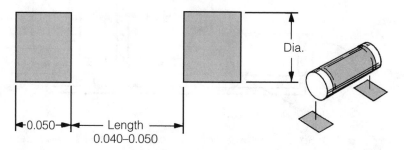

■ **3-11** *Land pattern geometry recommended for MELF devices is suitable for wave- and reflow-solder processes.*

Diodes and resistors can be supplied in the MELF package and generally maintain uniform physical tolerances. The end-cap termination plating on these components, however, is not always consistent from one manufacturer to another. Careful selection of the component supplier helps to reduce the necessity of artwork changes or process variation.

When reflow-solder processes are used, the MELF device might shift occasionally off the center line. For that reason, adhesive epoxy, as used in wave solder, can be applied to retain the component on the component side as an alternative. The notch pattern in the geometry provides a slight depression in the paste to nest the round device, but most likely, when the paste is the proper viscosity and thickness, the epoxies and the notch can be eliminated.

Plastic diode land pattern

Two package types have been developed as a preferred alternative to MELF packages. One has a gull wing lead similar to the SOIC design and the other type has a wrap under termination typical of those used for tantalum capacitors.

The smaller package of the two configurations is for general-purpose applications and the larger package accommodates the high-power-rated device types. The molded body provides a uniform flat surface that is more compatible with automated placement systems.

Tantalum capacitor land pattern

For many years the size for surface mount tantalum devices was determined by the factor of dielectric mass and working voltage. Because each supplier had unique body dimensions, multiple sourcing was almost impossible. The EIA membership of capacitor manufacturers focused on defining four uniform case sizes to package what they considered the most widely used capacitance range and an additional four sizes to accommodate the extended range capacitor.

There are subtle differences in these devices. It is important to select the value and power requirement early in your design because a change might require a significant redesign to accommodate a different case size and land pattern.

Molded devices such as tantalum capacitors and inductors are currently available from multiple sources. Careful selection of a reliable supplier for tantalum capacitors should be established early in the design planning.

Compare the choice with the standard tantalum configuration and value/voltage range in chapter 2. The contact geometry illustrated in Fig. 3-12 works well with the EIA standard tantalum capacitor in A, B, C, and D case sizes.

Inductor land patterns

Standards for surface mount inductor packaging are not as defined as most of the discrete device families. Depending on the

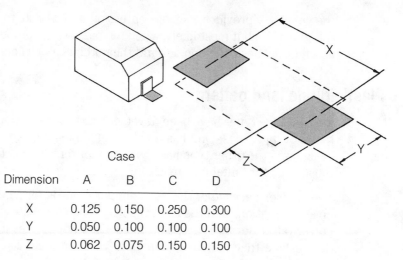

Dimension	Case			
	A	B	C	D
X	0.125	0.150	0.250	0.300
Y	0.050	0.100	0.100	0.100
Z	0.062	0.075	0.150	0.150

■ **3-12** *Land pattern width should be very close to the component contact width. Excessive area of the patterns will cause component shift during the reflow-solder processing.*

manufacturer, materials, and manufacturing techniques, these devices can have very unique physical characteristics.

The land patterns developed for inductor package sizes listed do not reflect the value of a specific device. The designer might find that a particular value might be available from more than one source in the same package. However, there is no industry-wide standard that has been agreed to by all suppliers. Other inductor configurations are available, but because of their widely different physical differences, the designer should rely on the specific data sheets supplied by the manufacturer for developing land pattern geometry.

Transistor land patterns

Several package styles are used for transistor devices, but the most common for general-purpose commercial application is the SOT-23 (or EIA TO-236). Although widely used for transistors, this three-leaded package style is also used for many single and dual diode types. Generally the SOT-23 package will adhere to a defined physical standard, but the tolerances given to the manufactures are quite liberal.

SOT-23 contact geometry

The land pattern used for SOT-23 includes a pad for each of the three legs of the device. The SOT-23 patterns shown in Fig. 3-13 details one pattern recommended by the manufacturers and because of the extension of the pad toward the outside, is also suited for wave-solder applications. The other SOT-23 pattern developed for reflow solder has an extra long land pattern on only one side.

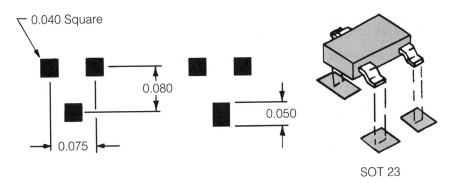

■ 3-13 *The three-point SOT-23 land pattern geometry of equal pad size will generally work for reflow or wave solder. Extending the contact of the single offset pad will reduce the occurrence of lifting during reflow processing.*

The extra long geometry on the center pin compensates for the imbalance of solder tension during the reflow process. Because the solder does not cool equally on all three pads, the side with two contact areas has the tendency to dominate. Without this longer geometry to offset contact, the component might tend to lift away from the board surface on one side as the solder cools.

SOT-23 placement and spacing

Many of the clearance rules used for chip components will apply to SOT components as well. When arranging SOT components on the PC board, the designer must provide enough physical clearance to allow for the placement accuracy of the assembly equipment being used. The guidelines for contact area to via hole pad distance recommended for chip components are also valid for SOT devices, as shown in Fig. 3-14.

An important factor to be noted is that the designer might make slight adjustments to adapt the contact locations to meet a desired grid location.

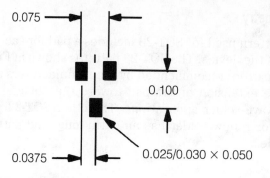

0.075

0.100

0.0375

0.025/0.030 × 0.050

■ **3-14** *One common rectangular land pattern geometry for all three SOT-23 contact locations is an option that may be considered.*

SOT-89 contact geometry

The SOT-89 package will accommodate the larger transistor die sizes and slightly higher power operation of some devices. The large area of the center tab aids in dissipating heat away from the component body to the surface of the board. The land pattern dimensions shown in Fig. 3-15 do not lend themselves to conventional grid position, but they are recommended by the component manufacturer.

Due to process and manufacturing yield problems experienced by both suppliers and users, this package is not recommended for

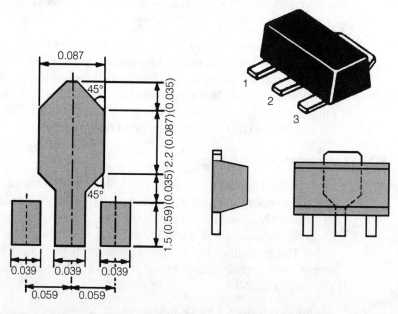

0.087

45°

45°

1.5 (0.59)(0.035)(0.087)(0.035)

2.2

0.039   0.039   0.039

0.059   0.059

1
2
3

■ **3-15** *Land pattern geometry for the SOT-89 power transistor provides a larger area of solder contact at the collector to improve heat transfer.*

new designs. Other packages better suited for power transistors include a modified TO-220 and a package developed specifically to replace the SOT-89, which is the SOT-223.

SOT-143

The SOT-143 is a four-leaded device that is packaged in a case size that is physically very much like the SOT-23. The four-lead configuration allows the component manufacturers an alternative for multiple diode and Field Effect or Darlington transistors. To assist in identifying orientation of the device, Fig. 3-16 identifies the larger contact of the four as pin #1.

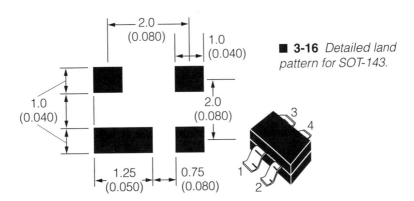

■ **3-16** *Detailed land pattern for SOT-143.*

A similar configuration is available from EIAJ manufacturers. However, the designer will find that they are larger overall and the wide lead is in a different location.

Small outline IC

The narrow SO-8, SO-14, and SO-16 ICs are generally used for packaging the smaller die sizes. The lead or pin assignment of the SOIC is usually the same as the larger DIP package, but occasionally they will differ and it is advisable to check the manufacturer's specifications carefully before beginning the design. Some devices require a larger lead frame to accommodate the die size. This wide format, referred to as the SO-L, SO-W, or SO-X extends from 8 leads up to 36 leads (SO-X). See Fig. 3-17. The SOP (EIAJ) family of devices, generally wider than the JEDEC SOIC packages, includes 8-lead through 42-lead package standards.

A pattern similar to the one shown in Fig. 3-18 can be used as a solution to the second source problem when the same function is

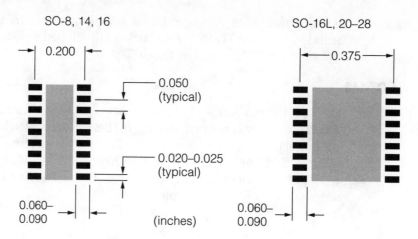

SO-8, 14, 16   SO-16L, 20–28

0.200

0.050
(typical)

0.020–0.025
(typical)

0.375

0.060–
0.090

(inches)

0.060–
0.090

■ **3-17** *The land patterns furnished for the JEDEC-registered SOIC family of devices conform to the IPC-SM 782 standards.*

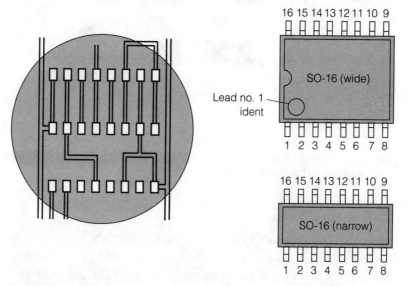

16 15 14 13 12 11 10 9

SO-16 (wide)

Lead no. 1
ident

1 2 3 4 5 6 7 8

16 15 14 13 12 11 10 9

SO-16 (narrow)

1 2 3 4 5 6 7 8

■ **3-18** *On rare occasions, a land pattern must accommodate a wide and narrow SOIC for a device with the same function from two different sources.*

available in the narrow package from Vendor A, while Vendor B supplies only the wide package. The dual row geometry will use more surface area on the board and might not be a practical solution for all situations.

SOIC contact geometry

The JEDEC-registered small outline, or SOIC contact pattern, is easily adaptable to CAD layout. For most logic devices, the narrow-body, small-outline configuration with 1.27-mm (0.050-in) lead pitch will have a 5.0-mm (0.200-in) span between contact-row centers.

Some 16-pin devices, however, require a wider lead frame to provide for the die size, or occasionally heat dissipation. A design might require a mix of the SO-16N and SO-16L (wide) devices. SOICs greater than 16 leads will always be one of the wider lead frames noted. Active components, beyond 28 leads, might also be available in the PLCC or other quad configurations or fine-pitch, gull-wing SO configuration. (See Fig. 3-19.)

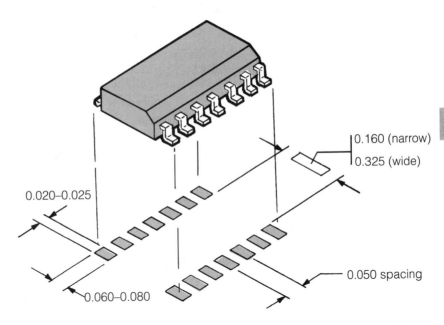

0.160 (narrow)
0.325 (wide)

0.020–0.025

0.050 spacing

0.060–0.080

■ **3-19** *The land pattern recommended for the JEDEC-standard SOIC package family will allow the component to float and self-center itself during reflow.*

Land pattern for the EIAJ SOP family

Active components in the EIAJ package can be obtained from multiple sources in Japan and are pin-for-pin compatible with many devices manufactured in the JEDEC format. The first suppliers of the SOIC chose the 0.050-in lead spacing to be compatible with the accepted United States "inch-based" grid system.

While the international committee of component manufacturers uses the metric system for developing devices, they do maintain the 0.050-in spacing for lead pitch on the SOIC packages. But among products defined by the EIAJ with 0.050-in lead spacing, there are significant mechanical differences. The most significant difference is the distance between rows of contacts. The EIAJ-SOP IC physically is not the same as on the JEDEC-SOIC. It is wise to study the physical specifications carefully on offshore-supplied devices, because they often vary greatly from one manufacturer to another.

If it is necessary to use devices from offshore sources, it might not be possible to mount both offshore and domestic sizes using the specific land pattern recommended by the domestic sources.

Alternative land patterns for JEDEC/EIAJ devices

SOICs are widely available from many EIAJ member suppliers and when these parts are to be used as an alternative to JEDEC SOIC's, a unique land pattern must be provided to ensure interchange or substitution. Even though the overall size of the device is different, pad geometry can often be adjusted to allow substitution on JEDEC and EIAJ SOICs. By lengthening the contact area, a wider part can be mounted to the PC board with minimal impact on the assembly process. Figure 3-20 illustrates a typical solder connection to the PC board.

Because of the excess length of the universal contact area, the narrower IC might tend to shift to one side during the solder reflow process. If component sourcing strategy requires both onshore and offshore suppliers, then the SO/SOP universal pattern is a configuration that should be considered. If the slight shift during reflow soldering is not acceptable, adhesive can be added under the component when the IC is placed on the board surface. Many EIAJ component suppliers do comply with JEDEC standard as well and they offer direct physical compatibility with JEDEC-defined devices.

Other two-sided gull-wing style surface mount devices are available as well. Families of low-profile, fine-pitch small-outline configurations are in production that follow industry-agreed-upon standards. One standard is defined as a Thin Small Outline Package (TSOP), and although the gull-wing-shaped leads only protrude at two edges of the device, the contacts are at the ends of the package rather than the sides.

62

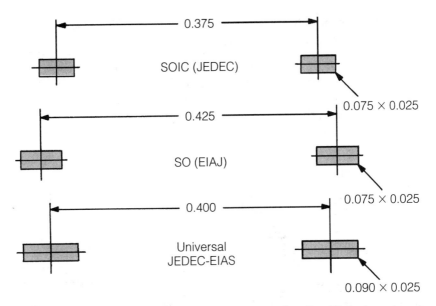

0.375

SOIC (JEDEC)

0.075 × 0.025

0.425

SO (EIAJ)

0.075 × 0.025

0.400

Universal
JEDEC-EIAS

0.090 × 0.025

■ **3-20** *Because many offshore sources for small outline IC devices do not conform to JEDEC standards, it is often necessary to adopt a universal land pattern.*

TSOP land patterns

The land patterns offered for the fine-pitch TSOP will vary in both physical geometry and lead pitch. Lead pitch is generally determined by the number of contacts needed for a particular device's function, while the overall package size is governed by the die size requirements.

Two manufacturers might offer the same device function, which is pin-for-pin compatible, and while the devices are different in body lengths. The designer might prefer the smaller device because of board space restrictions. However, provisions might need to be made for a second source that will require accommodating both narrow and wide package sizes typical of that detailed for the SOIC.

SOJ device

A dual inline configuration common for both JEDEC and EIAJ is the SOJ-IC. Originally, this package was introduced for memory products with parallel address and data lines. More manufacturers have found it to be a good alternative (in some cases) to the SOIC. The SOJ package has a larger mass that is better for heat dissipation or to accommodate larger die sizes. The SOJ IC is available in four standard body widths.

The land pattern options shown in Fig. 3-21 illustrate a standard geometry and an optional oval shape, which might best accommodate the minimum spacing of the JEDEC (SIMM) memory modules. Although the land area of the closely spaced ICs is reduced, adequate solder volume must be provided for a reliable connection.

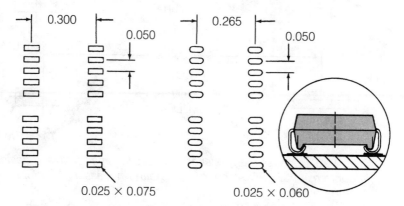

■ **3-21** *The detail compares two land pattern configurations for the .300-in wide SOJ. The wider pattern furnishes a more robust solder connection while the narrow pattern may be considered when devices have closer spacing typical for SIMM applications.*

Both SOJ land pattern options promote self-centering during the reflow process, and both shapes are compatible with reflow-solder processes.

Plastic-leaded chip carriers (PLCC)

The PLCC package proceeded the SOJ package design in using the J-lead contact. The unique shape gave the four-sided lead configuration a compliant contact to absorb the thermal mismatch between component material and substrate, thereby reducing stress on the solder connection.

Although available in lead counts as low as 18, the J-lead package is very popular for the PAL family of products and custom or semi-custom ICs numbering up to 84 leads.

The JEDEC terminal count column in Fig. 3-22 extends to 124 leads; however, the 68- and 84-lead devices are generally the practical limit due to the internal wire bonding length of the die from lead-frame, difficulty in maintaining lead coplanarity, and machine assembly handling.

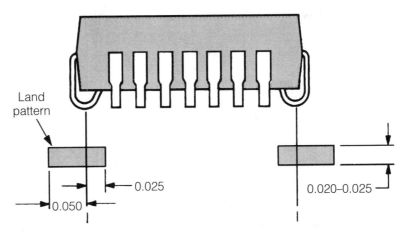

■ **3-22** *The PLCC land pattern is offset from the center-line of the J-lead to provide for a uniform solder fillet and easy visual inspection.*

The PLCC land pattern is not centered on the lead bend radius, but it is offset from the center line of the J-lead to provide for a uniform solder fillet to the outside heel area accommodating easy visual inspections. PLCC devices can be positioned with a minimum of 0.65 mm (0.025 in) between land pattern contact rows, but for visual inspection of PLCC solder joints and access for rework or repair, allow 3.8 mm (0.150 in) or more space between adjacent component leads. (See Fig. 3-23.)

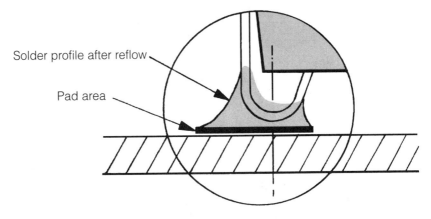

■ **3-23** *The solder fillet profile for the J lead package is influenced by both land pattern geometry and its relationship to the device lead profile or shape.*

The ideal PLCC land pattern should provide a good solder fillet toward the outside of the IC package lead. Extending the land pattern inward only increases the chance of component drift during the reflow process. Extending the land pattern's length even further to the outside for test-probe contact is only permitted if isolated by a narrow connecting trace or solder mask. Test probe contact on the primary side of the PC board will reduce usable board area for circuit routing and restrict component density. See Fig. 3-24 for details.

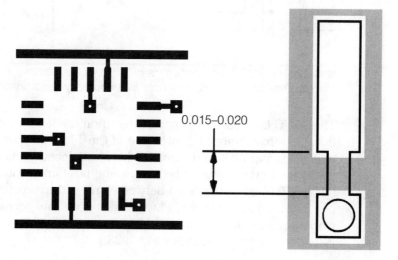

0.015–0.020

■ **3-24** *The space between land pattern and via hole and pad features must allow for a solder mask barrier to restrict the solder to the contact area.*

The distance between PLCC contact groups is usually determined by the complexity or density of the circuit. When bussing common signal traces on a multilayer PC board, a greater number of contacts of ICs can be interconnected on internal layers, as shown in Fig. 3-25.

All land patterns for the PLCC devices will allow the designer the option of routing a circuit trace between contacts. (See Fig. 3-26.) The solder fillet of the PLCC is controlled to a great extent by the land pattern geometry. The 0.63-mm (0.025-in) space between the land pattern will accommodate a 0.20-mm (0.008-in) wide circuit trace while maintaining an equal air gap at both sides.

The patterns shown in Fig. 3-27 accommodate the most widely used JEDEC PLCC devices. The land pattern for the PLCC should

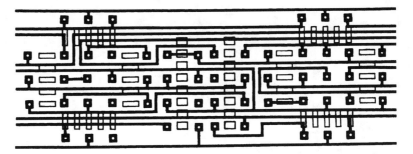

■ **3-25** *An SMT land pattern breakout to via hole and pad will increase component density on the substrate surface. All conductor trace interconnection would be shifted to internal layers for efficient CAD auto-routing.*

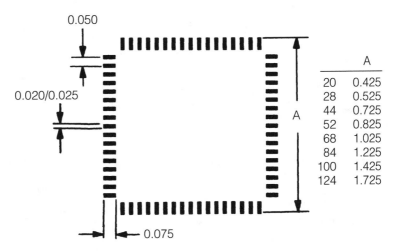

	A
20	0.425
28	0.525
44	0.725
52	0.825
68	1.025
84	1.225
100	1.425
124	1.725

■ **3-26** *Land pattern geometry recommended for the JEDEC standard square plastic leaded chip carrier (PLCC) device family.*

be separated from a via or feed-through hole by a narrow connecting trace.

Quad Flat Pack ICs

Quad-lead flat packs are one of the more popular surface mount packages for custom and semicustom ICs. Package size and contact spacing varies from one manufacturer to another, and could range anywhere from 1.27 mm, 1.0 mm, 0.8 mm, 0.63 mm, 0.5 mm (0.050 in, 0.040 in, 0.031 in, 0.025 in, 0.020 in) or less.

Because the fine-pitch Quad Flat Pack IC leads are more susceptible to damage and more difficult to handle in production, the de-

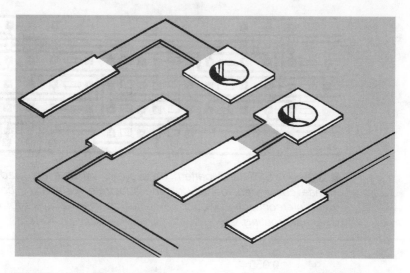

■ **3-27** *Via hole and pad for interconnecting the land pattern to other layers of the substrate must be separated by a narrow circuit trace.*

vices require special assembly-system capability. The closely spaced leads are very fragile and require special carriers, precise machine placement, and a refined soldering process.

JEDEC QFP land pattern

The JEDEC QFP device was developed by United States' manufacturers using inch measurements rather than the international metric system. The dimensions offered for land pattern development are given in "hard-inch" numbers with metric equivalents noted in parentheses (mm). Six-package sizing having 0.025-in (0.63-mm) lead pitch sizes are defined by the standard having a maximum allowance for 244 leads within a 1.6-in (40.6-mm) square transfer molded plastic housing. Land pattern geometry noted in Fig. 3-28 is conservative and should furnish the opportunity for a very uniform and reliable solder connection.

The plastic housing of these devices might include optional corner barrier features. When positioning other passive devices around the perimeter of the JEDEC QFP, the corner zones should be avoided.

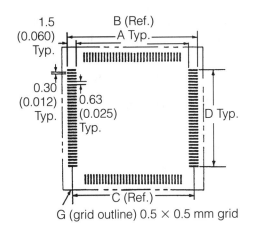

G (grid outline) 0.5 × 0.5 mm grid

Terminal count	A	B	C	D	G
84	17.5 (0.690)	20.5 (0.810)	19.0 (0.750)	12.7 (0.500)	42 × 42
100	20.0 (0.785)	23.0 (0.905)	21.5 (0.845)	15.24 (0.600)	48 × 48
132*	25.0 (0.985)	28.0 (1.105)	26.5 (1.045)	20.32 (0.800)	58 × 58
164	30.2 (1.190)	33.2 (1.310)	31.7 (1.250)	25.4 (1.0)	68 × 68
196	35.2 (1.385)	38.2 (1.505)	36.7 (1.445)	30.48 (1.20)	78 × 78
244	39.5 (1.555)	42.5 (1.675)	41.0 (1.615)	38.10 (1.50)	86 × 86

■ **3-28** *This land pattern geometry is conservative and should furnish the opportunity for a very uniform and reliable solder connection.*

Ceramic IC devices for SMT

When a product must withstand extreme operating environments, it might be necessary to specify ICs in a ceramic package. Although they do not have the wide choices for package styles typical of commercial plastic products, there are several leadless and leaded ceramic configuration options including: Leadless ceramic chip carrier (LCC), Ceramic quad with J-bend leads (LDCC), Ceramic flat pack (FP), and Ceramic quad flat pack (QFP).

Ceramic packages typical of those shown in Fig. 3-29 are furnished for lower volume hi-rel applications, but can usually be handled by automated assembly systems. Because of the difference in thermal coefficient of expansion (TCE) between the ceramic package and the glass epoxy printed circuit board, leadless devices must be mounted to a substrate material more stable than the commercial-epoxy glass substrate. The excessive flexing of the board during temperature excursions will stress the solder connection and eventually cause failure.

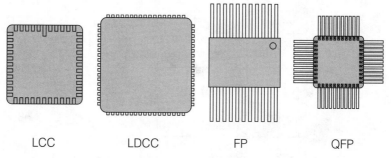

LCC   LDCC   FP   QFP

■ **3-29** *Ceramic package ICs are available in several lead configurations. Specify device leads trimmed and preformed for direct mounting when possible.*

Leaded ceramic package ICs are generally preferred and available in several configurations, but when possible, specify device leads pre-formed and tinned from the supplier for direct mounting. For example, ceramic components (FP) with excessively long gold-plated leads extending straight from the sides require extensive modification before being attached to the substrate (Fig. 3-30). The advantage of using the ceramic products with leads is the compatibility with a large variety of substrate types.

Unlike the leadless ceramic devices, the leads will flex sufficiently during the temperature variables, acting as shock absorbers for the solder bond.

Pin #1 on the contact area of the ceramic LCC leadless device is longer than the others for easy identification. If the mating contact on the substrate is also lengthened, it will act as a visual orientation guide and thus avoid the danger of unwanted interconnection

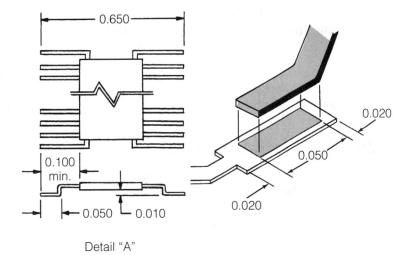

0.650

0.100 min.

0.050 — 0.010

0.020

0.050

0.020

Detail "A"

■ **3-30** *The land pattern recommended for ceramic IC device attachment will meet the criteria outlined in the industry developed Solder Joint Requirements Standard, J-STD--001.*

to vias or circuit traces that might be too near the extended contact.

Ball grid array land patterns

Land pattern geometry for BGA device attachment is round in shape and spaced in a uniform grid matrix. Three lead pitch standards established are 1.0 mm (0.040 in), 1.27 mm (0.050 in) and 1.5 mm (0.060 in). The diameter of the attachment site is somewhat dependent on the size of the solder alloy ball contact used. A common sphere diameter furnished on the commercial plastic BGA package is 0.6 to 0.8 mm (0.023 to 0.031 in). Via pad and hole patterns must be developed as a part of the attachment pattern. The via is typically positioned in the clear zone between the device attachment lands, connected with a short narrow signal trace, and covered by solder mask material.

Refer to the BGA package descriptions in chapter 2 for specific device size and contact pattern data.

Die Grid Array land patterns

Attachment of Die Grid Array devices will require a more refined land pattern geometry than the BGA noted above. The DGA package does have an alloy shere or bump contact for attachment, but the space between the contacts are 0.5 mm (0.020 in) or less. Mounting the IC package with solder is a preferred option but conductive adhesives can also be employed. In either case, the attachment site diameter will be only 0.05 mm and will not allow for a separated via pad for signal routing.

Array patterns for the DGA component family are device dependent and because each die size will have a slightly different I/O, each will vary a great deal from one application to the other.

Requirements for CAD/CAM data transfer

As the complexity of electronic products has increased, it is apparent the industry must utilize the efficiency of assembly automation as effectively as possible. Driven by the competitive demands to meet higher performance and reduced size for electronic products, engineers and designers often adapt several assembly methods. With the predominant usage of surface mount for PC board products, the design and materials specified for the board must be matched to diverse component and assembly-process technologies.

The more sophisticated assembly machines use automated vision systems for precise alignment adjustment in solder-paste application and device placement. To accommodate the vision features, the designer must provide fiducial targets on the board surface. The target is etched into the board to minimize dimensional tolerance accumulation to the land pattern geometry provided for device attachment. Three global fiducial targets are required for each board side having surface mount devices. Two additional fiducial targets for each fine-pitch device is recommended as well. The standard (minimum) size of the target is (0.040-in) 1.0-mm diameter with a clearance of 0.5 mm, void of solder mask and other physical features as shown in Fig. 3-31.

The standard fiducial is a solid-filled etched copper circle (0.040 in) 1.0 mm in diameter. A clear zone must be provided around the fiducial diameter and devoid of any other circuit features as well as solder mask material.

Establishing a component data base for CAD/CAM

Prior to designing the PC board in the CAD system, each component is constructed in digital form creating an electronic data base. The CAD data is most often used to prepare photo-tool artwork, printed circuit board fabrication details, and assembly instructions; however, if developed in the correct format, it can also be adapted to manufacturing processes. Direct transfer of CAD

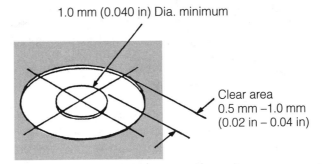

1.0 mm (0.040 in) Dia. minimum

Clear area
0.5 mm –1.0 mm
(0.02 in – 0.04 in)

■ **3-31** *The standard (minimum) size of the fiducial target is 0.040-in (1.0-mm) diameter with a clearance of 0.5 mm.*

data into automated assembly systems will accelerate production set-up and reduce overall assembly equipment programming time.

When the CAD data base is formatted correctly, specific physical data for each device can be used to program component placement (X-Y coordinate position) and orientation. To facilitate the X-Y coordinate information, a datum position must be established on the board surface. The recommended datum "0" for X and Y coordinates, is one of the global fiducial targets at the lower-left or lower-right corner of the board or panel, typical of that shown in Fig. 3-32.

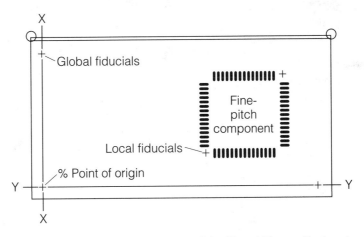

■ **3-32** *The recommended datum "0" for X and Y coordinates, is one of the global fiducial targets at the lower-left or lower-right corner of the board or panel.*

Review of component packaging standards

Surface mount devices are furnished in tape and reel as well as tube magazine feeders to accommodate high-speed assembly systems (tray feeders are most often adapted for fine-pitch components).

Each surface mount device is aligned using the body center and a starting orientation for reference. The basic orientation of the device is "0" degree. Rotational data must be specified from the "0" position in a counter-clockwise direction (typically 90°–180°–270°). The "0" starting position of the component is significant. See details in Fig. 3-33.

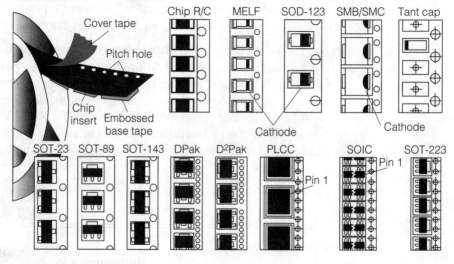

■ **3-33** *Standard device/packaging orientation.*

Tape-and-reel packaged devices for example, have an established standard for orientation related to the perforated pattern of the tape. The standard orientation does vary, however, between unique device families.

Orientation as well as polarity of a device must be defined in the CAD data base if the output transferred to the assembly system is to be reliable. Resistor and monolithic capacitor devices are common in orientation and have no defined polarity. As the designer develops the component data base, numbers are assigned to each end of the device to accommodate circuit routing and maintain orientation of value marking or polarity. Tantalum capacitors, diodes, ICs, and other polarized components, for example, have unique orientation in relation to tape-feed systems. In all cases, keep in mind the relationship of the device orientation within the tape cavity (see Fig. 3-34) in relation to the perforation at the tape carrier material edge.

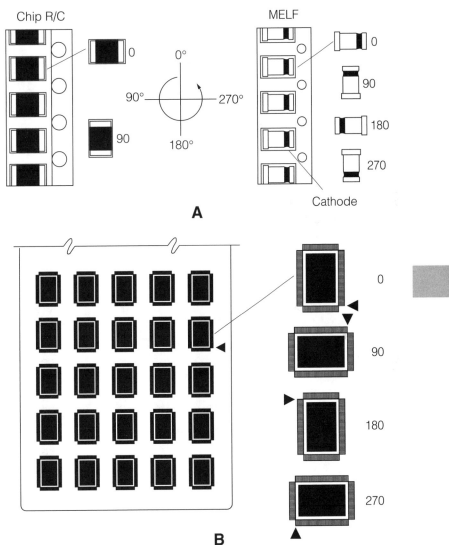

■ **3-34** *When developing the CAD data base, consider the basic device profile within the tape cavity or tray carrier as the zero orientation position.*

Devices supplied in tube magazine carriers and flat tray carriers will not be consistent with the tape-and-reel packaging system. For device types supplied in either tube or tray, machine programmers recommend that the designer maintain the orientation established for the tape-and-reel packaged component. Surface mount devices packaged in flat-pack tray carriers are usually not suited for tape-and-reel, except for the odd configurations (transformers, multichip modules, connectors). Most devices furnished in the tray format are the delicate fine-pitch product because the partitioned tray is specifically designed to restrict movement of the devices in transit, protecting the leads from deforming.

Developing usable machine program data

To facilitate automated machine programming, the position data for each surface mount device should be furnished in hard-copy format or disc file. The data defines the part number, reference designator, X - Y coordinate location, and orientation for each device on the assembly, as shown in Fig. 3-35. Data for machine programming can be developed by other methods, but the manual programming process might require one or two days to complete and verify accuracy.

Device ref. desig.	Part number	Location X	Y	Device orientation	Device type
C512	305002401	1695	335	0	0603
C516	305001701	2115	705	90	0805
C517	305001701	1975	60	90	0805
C29	305004101	2429	456	270	1206
L5	315000401	1479	1094	0	1206
L9	315001101	1541	616	180	1206
R53	300000101	1381	1096	180	1206
R709	300000101	2096	149	90	1206
U9	330000601	1340	1419	90	4661
U8	330005801	1842	1530	180	4722
U10	330000501	1673	907	0	8020
Q4	320000701	−487	1618	180	9952
U11	330002601	123	431	0	AD-7180
U6	335000201	1027	200	180	ASIC
C509	305007801	−627	1754	270	ATANT
C510	305007801	2139	988	180	ATANT
C511	305007801	2002	794	270	ATANT
U5	330002501	1510	215	180	BA100
U17	320000601	1340	790	180	CA3083
C1	305003101	−500	1433	180	DTANT

■ **3-35** *The data defines the part number, reference designator, X - Y coordinate location, and orientation for each device on the assembly.*

CAD data for fixture development

The preparation of solder-paste stencils and other fixtures requires a disc copy of the Gerber data and aperture list developed for fabricating the PC board. Manufacturing engineers, for example, use the solder-paste layer(s) of the file to prepare stencil fixtures. The engineer modifies the land pattern geometries when necessary to regulate or control solder paste volume of specific device types. When stencil systems are equipped with vision assisted alignment of the solder stencil fixture to the PC board, the global fiducial target images furnished for automated component placement can also be merged onto the stencil pattern (especially when fine-pitch devices are in use).

77

Space planning and interface

Space study for SMT

THE PRINCIPAL GOALS OF THE PLANNING STAGE OF THE surface mount product are to define the inter-relationship between all devices as well as estimate adequate spacing or clearance needed to ensure an easy-to-manufacture assembly process. As component density increases on the circuit board, other issues that must be addressed include: reduced conductor width; additional circuit layers; inspection, testing, and rework issues, which become more difficult.

To assist the designer in the organization of components and estimation of the density factor on the board, a preplanning procedure or "space study" is recommended. See details in Fig. 4-1.

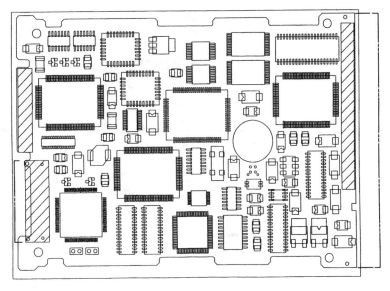

■ **4-1** *A sample board layout to assist the designer in the organization of components and estimation of the density factor on the board.*

Estimating total component area

The component body dimensions alone do not furnish sufficient information to determine board space. The overall dimensions of the component land pattern must be included in the preliminary planning of the board (Fig. 4-2). In addition, a percentage multiplier is added to the land pattern area to allow for conductor traces and via hole pads.

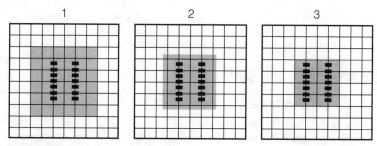

■ **4-2** *Density factor is measured by the space required to mount and interconnect the components. As the density factor compresses or space around each device is reduced, circuit layers must be added to accommodate circuit routing.*

Interconnection of dissimilar component types might require extending the space between devices. The SOICs shown in Fig. 4-3 require less surface area for routing circuits than the PLCC. Mixing the two component types on the board is common, but routing of traces between physically unique devices becomes more difficult. Choosing components that can make efficient use of space will best facilitate interconnection.

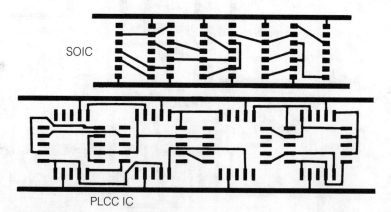

SOIC

PLCC IC

■ **4-3** *The density factor is affected dramatically by the package type selected for the circuit. Logic components in the SO package will require less space for routing circuit traces when compared to the PLCC IC interconnection.*

Component spacing for SMT

To maximize component density, designers might attempt to pack components on the PC board with only minimum clearance between component bodies. In surface mount, component spacing can impact the efficiency of automated assembly processing. Robotic assembly equipment places surface mount devices (SMDs) on the substrate surface with an accuracy within ±0.127 mm (0.005 in) to 0.025 mm (0.001 in) of true position. The recommended guidelines provide space between components for ease of inspection and/or rework.

Component placement

During the initial placement phase, the designer plans the optimum location of components for efficient interconnection of related devices. Interconnection of components with conductor paths on the same surface as the components present a challenge to the designer.

The arrangement of components in a functional relationship to each other is only one consideration in planning the circuit routing of the surface mount PC board. This interconnection factor requires careful attention to the relationship, orientation, and placement of the devices as shown in Fig. 4-4.

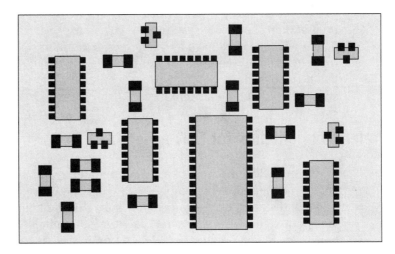

■ **4-4** *Direct coupling of devices on the substrate is often necessary for analog or RF circuits. The difficulty and additional labor required for assembly is generally an accepted cost associated with these products.*

The illustration shown in Fig. 4-5 is an example of positioning components for the most direct interconnection possible. While this direct coupling for analog or layout-sensitive circuits is not uncommon, it might use more surface area for routing circuit traces.

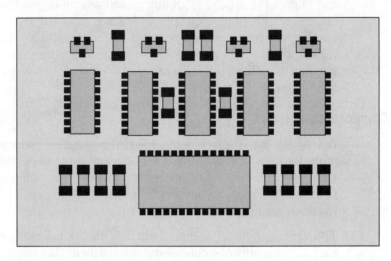

■ **4-5** *Common direction of component parts will increase assembly efficiency and reduce overall manufacturing cost.*

The direction of the SOT devices, ICs, and other chip component programming assembly systems and visual inspections is more efficient when devices have a common orientation or polarity. Even though assembly equipment will rotate each part before placement, the designer should plan the layout carefully to maintain a consistent component orientation.

Preferred orientation for SMT assembly

Common device orientation, as illustrated in Fig. 4-6 promotes efficient, cost-effective assembly. Both component orientation and signal paths are planned to take advantage of the board's surface area, although device placement is often determined by the relation to interface connectors. Related or interactive devices are grouped into functional clusters to make the most direct circuit trace interconnection possible.

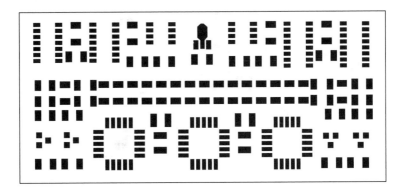

■ **4-6** *Organization of the component layout will simplify the assembly process and reduce manufacturing cost.*

Consistent orientation of components will expedite the assembly process, a cost-saving factor that must be considered during the design phase of the project.

Fine-pitch, QFP, and SOIC space planning

Fine-pitch devices with center-to-center lead spacing of 0.63, 0.50 mm (0.025, 0.020 in), or less, usually require a reduced volume of solder paste for reflow processing, and specific spacing between the contact area of the QFP and other fine-pitch devices, which must be maintained for both solder-paste application and rework.

To adjust the volume of solder on fine-pitch land patterns, engineers might adapt a step-stencil process. If the stepped or multi-level solder stencil is to be used, a minimum area of 3.2 mm (0.125 in) must be provided between the overall contact patterns of adjacent components. The detail shown in Fig. 4-7 is an example of a group of SMT components including a typical fine-pitch device.

The QFP ICs might be placed and reflow soldered at the same time as other surface mount devices or attached as a secondary, single-station assembly process. The post-assembly operation might include a more precise placement system or a unique reflow-solder process. A detailed description of solder stencil fabrication and assembly options is included in chapters 7 and 10.

Using both sides of the substrate

Unless all components chosen for the assembly are available in a surface mount configuration, mixing surface mount components

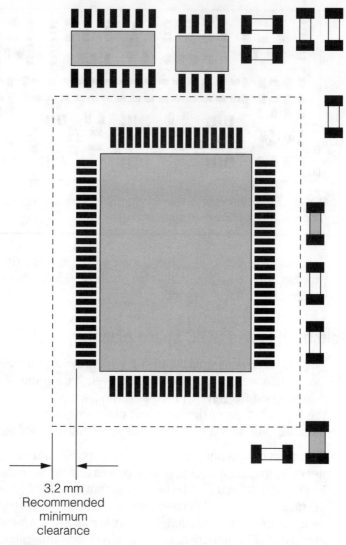

3.2 mm
Recommended
minimum
clearance

■ **4-7** *Provide a clear zone on all sides of the QFP device to allow for inspection rework and repair without disturbing adjacent devices mounted in the same area.*

with leaded devices will be unavoidable. One option open to the designer is the two-sided assembly. The majority of the active components are mounted on one side, while chip components are mounted on the secondary (wave solder) side. (See Fig. 4-8.)

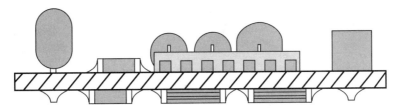

■ **4-8** *The mixed technology assembly example shown allows for a single wave solder process for attachment of both PIH devices and surface mount devices.*

Trace-to-contact guidelines for SMT

PC boards designed for SMT application have several assembly options dependent upon component density and interconnection complexity. Boards range from simple one- or two-circuit sides to the more complex multilayer-laminated construction common in high-density, high-performance assemblies. The trace width as well as the air gap between traces and pads will affect quality and cost of the finished board. Recommendations for circuit trace and air gap between traces for outer surfaces of a substrate should be no less than 0.20 mm (0.008 in.) When multilayer construction is required, fabricators will generally use either one ounce, 0.35 mm (0.0014-in thick), or one-half ounce 0.018 in (0.0007-in thick) copper-clad laminate. The one-half ounce copper on inner layers of the circuit board will allow for 0.13–0.15 mm (0.005–0.006 in) circuit trace width with equal airspace between traces.

The inner layers, using one-half ounce copper, can have trace width/air gap dimensions of 0.15 mm (0.006 in) and less; but, 0.20 mm (0.008 in) on the outside layers will have a higher yield and not cause excessive fabrication cost.

Trace width of 0.08 mm (0.003 in) with equal air gap might be an everyday occurrence for some fabricators, but remember, high-tech boards are closely associated with high cost. The mainstream quality shop doing multilayer PC boards will recommend a more conservative approach whenever possible to maximize yield and control costs.

Trace-to-trace and trace-to-contact patterns shown in Fig. 4-9 allow for consistent etch clearance on outer layers, while ensuring thorough coverage by the solder mask over the circuit trace.

An exposed edge of the conductor passing near or between contact areas will attract solder particles and result in bridging. Re-

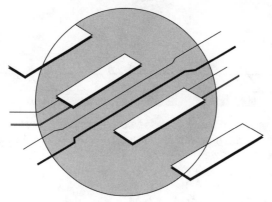

■ **4-9** *To prevent the possibility of solder bridging between circuit features, solder mask must completely cover the circuit traces that are routed near or between the device land patterns.*

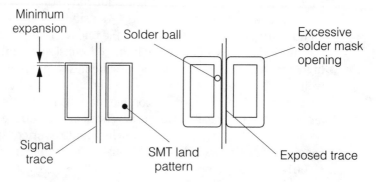

■ **4-10** *When solder alloy bridges between the land pattern and a non related circuit feature under the device body, the device must be removed to correct the defect.*

moval of the component is often the only way to eliminate solder bridging when it occurs under the component. See Fig. 4-10 for details.

Routing a 0.20-mm (0.008-in) conductor between 1.27-mm (0.050-in) spaced land patterns is a common design practice. However, the circuit trace must be covered with solder mask. The use of narrow traces to interconnect the wider conductor to the component contact area allows the solder to liquefy evenly with other land patterns.

If a very wide trace or copper area is connected to the contact pad, two reactions are possible during reflow solder. First, the component could be drawn off the liquid solder, because the solder contacts adjoining the large metal masses cool the solder quickly. In the case of chip components, one end of the component lifting away from the pad (commonly known as *tombstoning*) often occurs. The solder mask should be applied over bare copper traces

only. If the circuit traces are coated with solder plating, the lique-
fied solder paste will flow under the solder-mask coating and
merge with the plating on the conductor traces. If migration from
one or more leads or contacts of the surface mount device is excessive it might require manual touch-up.

Figure 4-11 illustrates a few "Dos and Don'ts" to avoid solder problems. Separation of the land patterns of two chip components will
ensure containment of the solder paste. When it is important to increase conductor width between components, the best result is
achieved with two narrow traces rather than one wide one. Containment of the solder within the footprint patterns is the key to
controlling the solder process.

Trace-to-contact guidelines

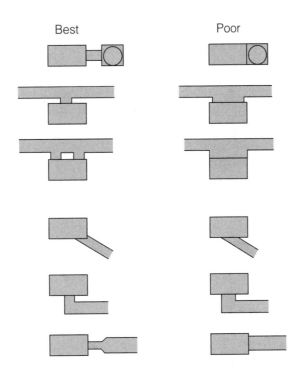

■ **4-11** *Guidelines for land pattern-to-conductor trace and via-to-con-
ductor trace interconnections have been well established. Adhering to
these guidelines will ensure a "process friendly" design.*

The preferred circuit board for surface mount assembly will have
solder mask over bare copper (SMOBC). The solder plating is limited to the exposed contact area of the components only while

conductor traces and other copper plating under the mask material remain flat. (See Fig. 4-12.) Plated through-holes (or via holes) and pad area must be separated from the land pattern by solder mask. The mask will contain the solder paste during the reflow-soldering process. (See Fig. 4-13.) Spacing between components that are connected end to end must be far enough apart to ensure solder mask separation between land patterns.

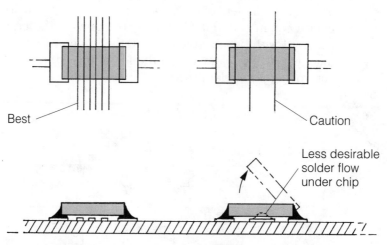

Best   Caution

Less desirable solder flow under chip

■ **4-12** *Exposed solder plated circuit traces under low profile surface mount devices cause serious assembly process complications and should be avoided.*

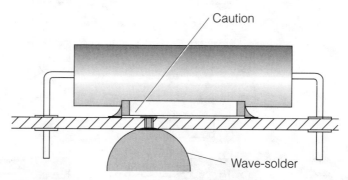

Caution

Wave-solder

■ **4-13** *If not covered by solder mask, liquid alloy can migrate through the small plated via during the wave solder process, causing damage or creating a bridge between via and land pattern.*

Spacing for higher profile components

The PC designer seeking maximum component density often mixes surface mount and leaded devices on the same substrate, but fails to provide adequate spacing between devices. Mixing high- and low-profile surface mount devices with leaded connectors or DIP ICs demands greater attention to specific clearance guidelines.

Spacing between the taller SMT components must also permit visual inspection of the solder connections. A viewing angle to inspect the solder connection of a J-lead Plastic Chip Carrier (PLCC), requires a spacing of no less than 3.8 mm (0.150 in). See Fig. 4-14. This component spacing also reserves access for solder touch-up or rework tools and test clip attachment. The same angle of view must be calculated when positioning taller molded capacitors or inductors near the high-profile ICs.

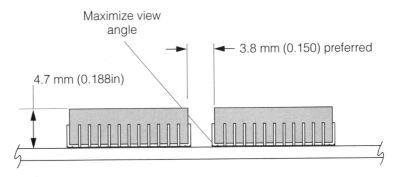

■ **4-14** *Spacing between the taller SMT components must also permit visual inspection of the solder connections.*

It is common to mount DIP ICs parallel to one another with 2.5-mm (0.100-in) spacing between lead rows. The solder connection to the lead- or pin-in-hole (PIH), detailed in Fig. 4-15, allows for solder inspection on the side opposite the component body. When mixing the SMT and PIH devices in close proximity, the designer should plan for a viewing angle and rework tool access equal to the space recommended for the PLCC previously noted.

Modular PC board planning

Mixed technology guidelines are expanded on later in this chapter. Single In-Line Memory Modules (SIMM) assemblies are limited to a specific profile. The density of the SMT assembly occasionally dictates closer spacing than previously recommended for PLCC and SOJ devices and can be mounted with a minimum clearance as shown in Fig. 4-16.

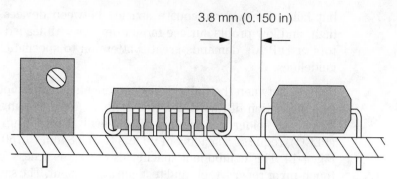

3.8 mm (0.150 in)

■ **4-15** *When locating higher profile SMT components near other devices, reserve space for solder inspection and, if necessary, touchup tools.*

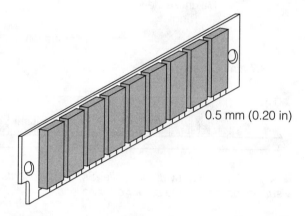

0.5 mm (0.20 in)

■ **4-16** *Because the size of the standard SIMM profile is restricted, these assemblies often force the designed to violate the component spaing recommended for general or system size printed circuit board applications.*

Some J-lead components must be spaced as closely as 0.5 mm (0.020 in) between lead surfaces. Visual inspection in this case is not possible and solder-process verification can be made only with electrical testing or x-ray evaluation. Both these methods require specialized equipment with skilled technicians, and generally x-ray inspection is on a sampling basis.

When the SMT component density exceeds that previously recommended, the surface area on the secondary side can be used. Although all surface mount device types can be attached to the secondary side, transferring the lower-profile discrete components reduces component density with very little effect on overall assembly cost. Chapter 5 illustrates the two-sided SMT assembly layout and clearance guidelines in greater detail.

Clearance between chip components must also accommodate inspection and rework tools. With adequate space, the danger of solder bridging and voids is eliminated. Figure 4-17 is a general guideline for component spacing for chip devices attached with the wave-solder process. The illustration includes the contact pattern and component body.

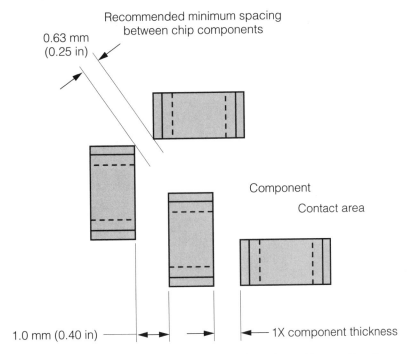

■ **4-17** *Chip resistors and capacitors can be oriented in any direction, but the spacing between components must remain accessible to repair or replacement tools.*

Wave solder processing surface mount devices

When leaded (PIH) components are mixed with the surface mount assembly, attachment of these devices is usually performed using a wave-solder process. By mounting the majority of the low-profile chip components on the wave-solder side, more surface area is reserved for trace interconnection on the component side. The low-profile chip components are generally located well below the finished length of the lead ends of PIH components. Resistors and capacitors are attached to the wave-solder side of the assembly by cured epoxy developed specifically for the secondary wave-solder process.

The chip component land patterns shown are recommended for reflow-solder processes, but also work well on single or dual wave-solder machinery. Because wave soldering is a mature process, companies might choose to use leaded through-hole components for all ICs and larger passive devices. By mounting the majority of surface mount resistors, capacitors, and transistors with epoxy adhesive on the wave-solder side, overall goals of board size reduction can be achieved. Mounting chip components on the wave-solder side, however, will restrict conductor routing paths. See Fig. 4-18 for details.

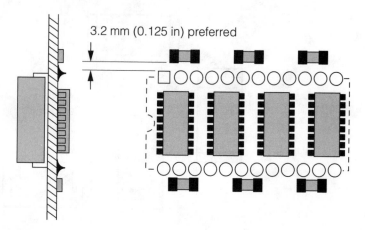

■ **4-18** *Wave solder of leaded parts often follows reflow processing of the surface mount components. Allow spacing between the device leads and SMT components for solder masking fixtures.*

The land patterns illustrated in Fig. 4-19 allow for the typical placement accuracy of assembly equipment, as well as process variables such as component shift during adhesive cure. When using the narrower contact pattern, be sure to allow adequate clearance between the component bodies, conductor traces, and vias. Keep in mind that the component body will overlap both sides of the land pattern when using this narrow design.

Many land pattern design possibilities exist that further improve process yield. Component land patterns continue to be refined to

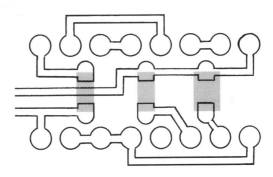

■ **4-19** *A narrow contact geometry will reduce excessive solder buildup on a component during wave-solder connections.*

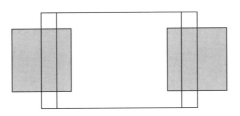

■ **4-20** *Although a full width land pattern works equally well for reflow and wave solder processes, the narrow land patttern options shown will restrict excessive solder buildup and physical stress on devices.*

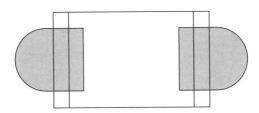

reduce or eliminate the need for rework and touch-up of boards after assembly processing. Figure 4-20 compares alternative contact geometry used for wave-solder application.

Careful planning of surface mount chip component placement during layout will take advantage of small spaces left unused by bulky leaded counterparts. By placing surface mount components in close proximity to related leaded devices, the circuit connection is more direct. Efficient use of this unused area might allow for more efficient use of valuable surface area on the primary side of the PC board.

Mixing leaded components on the SMT assembly adds steps to the assembly and soldering process, and more complex assembly processes involve the use of additional fixtures and assembly equipment.

In Fig. 4-21, the low-profile SMT devices will easily nest between the radius of large axial lead components to make maximum use of the substrate area and reduce circuit conductor length. Close coupling of chip components on the opposite side of the leaded device can reduce line inductance and increase component density.

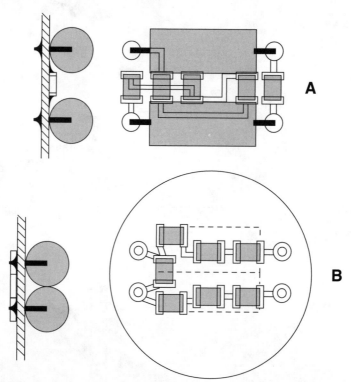

■ 4-21 *(A) The low profile SMT devices will easily nest between the radius of large axial lead components to make maximum use of the substrate area and reduce circuit conductor length. (B) Close coupling of chip components on the opposite side of the leaded device can reduce line inductance and increase component density.*

Locating surface mount ICs on the secondary side, as shown in Fig. 4-22, will maximize the surface area usage as well. Be aware of an increased occurrence of solder bridging between surface mount IC pins when wave soldered. If ample clearance is not reserved, touch-up will occasionally be required even with advanced dual wave-solder techniques.

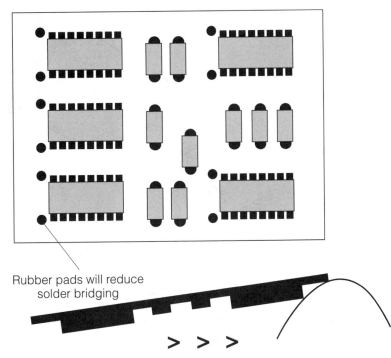

Rubber pads will reduce solder bridging

■ **4-22** *To avoid excessive solder bridging or other defects on surface mount devices during the wave solder process, mounting orientation must be considered.*

Connectors and interface for SMT assembly

Connector manufacturers have introduced several interface products for surface mount applications. In addition to the more prominent names in connectors, several specialty products from domestic and offshore suppliers are providing unique interconnection systems. Selecting a connector that will mate with existing connector or cable connector families is preferred, because this obviously provides flexibility and, in many cases, multiple sources (Fig. 4-23).

Some designers might not want to use a surface-attached connector because of the questionable physical strength of soldered contacts.

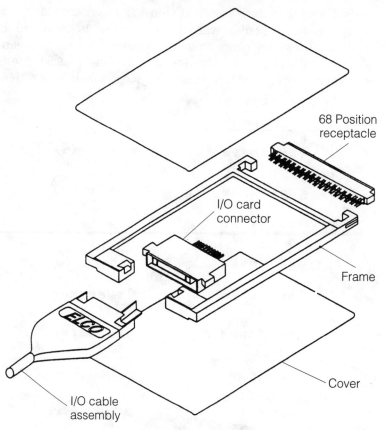

68 Position
receptacle

I/O card
connector

Frame

Cover

ELCO

I/O cable
assembly

■ **4-23** *PCM Card standards have very defined physical limitations that has enabled the development of a unique family or system of interface products.*

In some cases, connector manufacturers furnish retention or strain-relief mounting tabs, bosses, etc., for the user who requires additional mechanical support beyond the solder connection. Techniques for mechanically retaining the connector to the substrate surface will often vary from one manufacturer to another.

Heat seal

Another method of interfacing assemblies used in consumer products is the heat/pressure seal flat cable (HSC). A typical low-cost heat/pressure seal flexible cable is a polyester film-base material with parallel rows of graphite or graphite-silver conductors covered by an insulating layer. The ends of this flat cable are free of insulation. When heat and pressure are applied, the electrical connection is made to the substrate.

Heat seal is a popular technique for mating the PC board to the glass surface of the liquid crystal display (LCD) components. Refer to HSC flexible cable manufacturers for specifications on compatible mating contact materials and environmental limitations. Details are illustrated in Fig. 4-24.

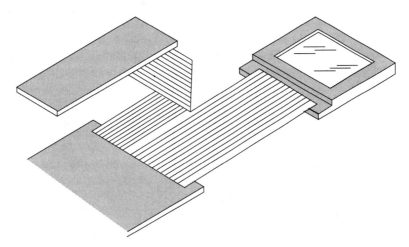

■ **4-24** *Flexible cable systems can be terminated to several substrate types without the need of a connector.*

Other conductive materials are available for interfacing flexible cables as well. The Z-axis conductive films, for example, have alloy particles suspended in close proximity to each other but will not make electrical connection between two surfaces until pressure and heat are applied. This material has proven to be a very reliable interface between the copper conductors on a polyimide-flexible circuit and the silver film contact typical of the liquid crystal display. When studying these alternatives for a specific application, the designer will find that manufacturers of the conductive materials are a good source for process instructions and reliability data.

Compression

Compression connectors are designed to join two or more parallel substrates that have matched contact areas on each mating surface as shown in Fig. 4-25. The substrates must be mechanically retained because this is usually how the connector material is captured. Connectors can be retained using common hardware or epoxy adhesive placed at the ends of the compression material.

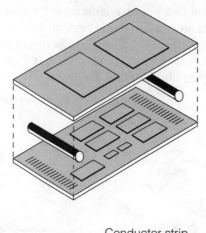

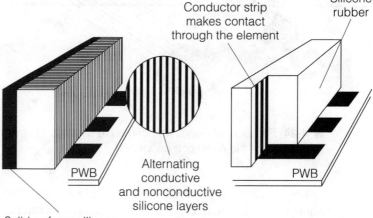

Conductor strip
makes contact
through the element

Silicone
rubber

PWB

Alternating
conductive
and nonconductive
silicone layers

PWB

Solid or foam silicone
for support or insulation

■ **4-25** *Mating parallel circuit assemblies with compression connectors will assure the lowest possible finished profile.*

Epoxy will permanently attach the connector to one substrate surface or the other.

DIP and SIP module design

The Dual In-line Package (DIP) module shown in Fig. 4-26 is a method of taking advantage of high-speed robotic assembly technology. Edge clip contacts are economical for direct soldering of DIP modules into the host substrate. The most common lead spacing is 2.5-mm (0.100-in) centers but a staggered 1.27-mm (0.050-in) spacing is also available for higher density applications.

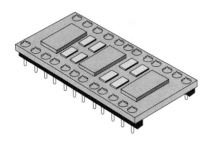

■ 4-26 *A Dual In-line Package (DIP) module.*

The size of the DIP module could conform to the limits of standard ICs or could take an outline adapted to individual needs. The designer is not limited to mounting components on one side of the module. It is a common practice in SMT, as well as in MCM assemblies, to use both surfaces to increase the density of the module.

Dual in-line contact strips are the most economical method of terminating this module. The contact strips work well when the module is mounted into assemblies that are to be wave soldered. The land pattern shown in Fig. 4-27 is for reflow or dip soldering the edge contacts to the module.

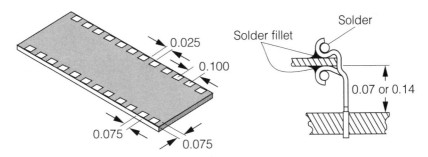

■ 4-27 *The footprint pattern is for reflow or dip soldering the edge contacts to the module.*

The edge-mount contacts are furnished in a continuous strip with a common breakaway bar to help retain alignment. Contacts can be supplied with solder preforms (solder alloys) built into the contact to ensure an even flow and a strong connection. Pin-and-socket strips or headers should be used for more durable contact requirements.

Typically, pin-and-socket connectors are used when a module needs to be added or replaced without special tools. The contact strips are best suited for modules requiring a low profile. The single in-line pin (SIP) module is popular for SMT.

Ceramic hybrid modules often are converted to laminated circuit board (MCM-L), thereby reducing substrate costs by 50 to 60 percent. It is common to partition the electronic functions into a module that can be used in many products. See Fig. 4-28. The SIP module is handled as a component, tested, and easily soldered into larger PC board assemblies.

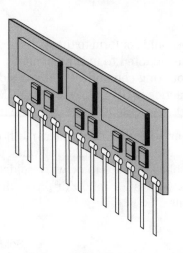

■ **4-28** *It is common to partition the electronic functions into a module that can be used in many products.*

Contacts for SIP modules are supplied plain or with solder to help the uniformity of the solder connection. SIP contacts can be mounted to the module prior to the solder-reflow process or added as a post-assembly procedure.

The land pattern shown in Fig. 4-29 provides excellent electrical and mechanical bonding characteristics after reflow-solder. When allowing for overall height of the SIP module, don't overlook the additional stand-off height built into the contact. Pin mounting is optional and must be specified by the user. It is important to study the manufacturer's specifications closely. Many of the contact pins are designed to mount into 2.5-mm (0.100-in) spaced hole patterns and sizes generally used for DIP ICs or other leaded devices.

The Single In-Line Pin (SIP) module assembly is very similar to the SIMM but because the SIM is soldered directly into the host board like a component, the extensions required for socket retention have been deleted.

Low profile edge pins

The edge clip pin configuration, although widely used for SIMM applications, requires an allowance for the interface forming be-

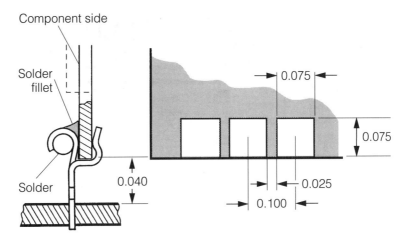

Component side

Solder fillet

Solder

0.040

0.075

0.075

0.025

0.100

■ **4-29** *SIP contact pattern.*

tween the edge of the module and the lead section designed to attach to the host PC board. A low-profile 2.54-mm (0.100-in) pitch pin design is available for attaching the SIM. The low-profile pin shown in Fig. 4-30 is designed to be hole-mounted and solder attached to the SIM assembly. The area reserved for mounting this pin style is smaller than the edge clip type and clearance between module edge, and the host PC board is minimized.

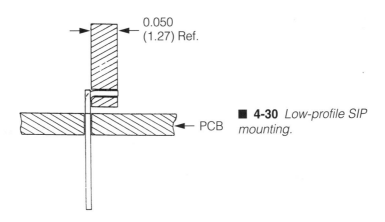

0.050
(1.27) Ref.

PCB

■ **4-30** *Low-profile SIP mounting.*

These pins will adapt the SMT module to a PIH PC board as a pretested component level part in the same way as the custom DIP module noted above. When the SOJ IC or other devices are mounted, as shown in Fig. 4-31, using minimum clearance between

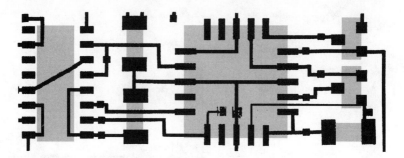

■ **4-31** *Because the area and size is limited, component placement and circuit routing on SIMM and SIP modules must be carefully planned.*

parts (typical of SIMM or SIP assemblies), trace routing on the primary component side can be restricted. The designer can maximize the limited space of the module by planning direct surface interconnection wherever possible and avoid excessive use of vias.

Standards for SMT memory module

DIP-style memory ICs are used on conventional PIH boards using sockets. PLCC, SOJ, or TSOP memory devices can be mounted directly to a PC board or furnished as an expansion or extension to the on-board memory. When space is at a premium, the designer must look at alternative packaging concepts. The miniature SMT ICs are difficult and expensive to socket as individual components, but as a modular plug-in set, users can add or expand memory functions without special tools or technical expertise.

The examples shown in Fig. 4-32 are JEDEC-registered module designs typical of those furnished by several suppliers. This standardization has assisted distributors and users of memory in maintaining several approved sources. SIMMs have several configurations, but PIH solder-mounted connectors have been developed to mate with all the standard modules.

Quad module design

Using the 1.27-mm (0.050-in) space contact leads on the perimeter of the module, the designer can develop a quad MCM-L using conventional PC board materials. Use the SMT carrier contact when possible and design the land pattern around one of the JEDEC standard arrangements. If the circuit is then converted to

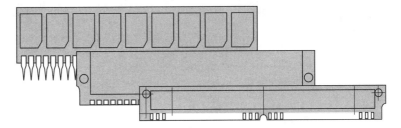

■ **4-32** *Single in-line memory modules (SIMM) were developed expressly for vertical mounting of surface mount assemblies.*

a single-chip plastic-molded device, the host PC board will not require modification.

Carrier contacts shown in Fig. 4-33 are supplied on a 1.27-mm (0.050-in) spacing and are ideally suited for miniature applications. The compliant contact design allows the mating of substrate materials and absorbs stress caused by different rates of expansion during thermal cycles.

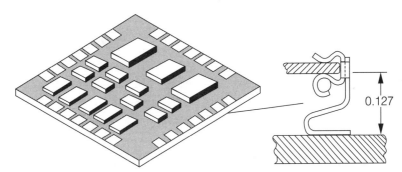

■ **4-33** *Carrier contacts are supplied on a 1.27-mm (0.050-in) spacing and are ideally suited for miniature applications.*

Further complex interfacing between a module and the host PC board is possible using design contacts having closer lead spacing. The quad lead design is often used to provide a working model to simulate a custom gate array or as an alternative for a product that is not presently available in a surface mount package from the component manufacturer. Using pretested standard bare IC chip packaging (MCM) and surface mount devices to configure the function of a custom IC might provide the advantage of a strategical early entry of a product into a highly competitive marketplace.

Layout guidelines for rigid circuits

THE SHAPE OF THE PRINTED CIRCUIT BOARD IS TYPICALLY established to meet specific aspects of the product. The PC board designer often must adapt the components and circuit to physical limits defined by an industry-standard format or to meet the unique shape defined by the product itself. Odd- or irregular-shaped boards are developed to conform to an enclosure designed for specific application or operating environment. System boards on the other hand generally are uniform in shape because they must be compatible with fixed card guides and mate with connectors or back planes.

The personal computer circuit boards are smaller than most system boards. Industry formats might include the PCMCIA, or Micro Channel standards, each having established physical features. Standards are established for PC or desk-top systems as far as connector interface or mounting features, but the overall length or width have liberal tolerances. Boards that are designed to be housed in nonstandard enclosures, typical of hand-held or lap-top electronics, are not under industry-defined limits. Because of the generally compact nature of these systems, the shape often must conform to a functional aspect of the product, keyboard, display interface, connectors, etc.

Although automation for SMT might be assumed, assembly of PC boards with unique shapes and odd sizes in low-volume or prototype quantities can be efficient. With proper fixturing, know-how, and physical resources (labor), skilled personnel can attach almost any SMT device using tweezers, vacuum tools, and a good soldering iron. Hand assembly is generally acceptable to a point but as more individuals are added to the assembly process, the varying levels of skill will often be reflected in the overall consistency of workmanship. As the quantity requirements increase, product quality might decrease.

Assembly automation for medium to high volume

Medium- to high-volume manufacturing of surface mount products requires some level of assembly process automation. Automated systems, typical of that shown in Fig. 5-1, are available for just about any phase in the assembly process. Solder printing systems for example, with or without vision alignment, pick and place for passive devices to fine-pitch and several systems for reflow or wave soldering are conveyorized for integration into an automated assembly process. Each specialized machine or system can function efficiently with manual transfer of the board through each process stage to accommodate low-volume requirements.

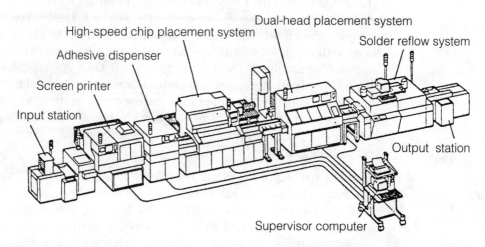

Dual-head placement system

High-speed chip placement system

Solder reflow system

Adhesive dispenser

Screen printer

Input station

Output station

Supervisor computer

■ **5-1** *An automated surface mount products assembly line.* Fuji America

PC board requirements for automation

Automated manufacturing or in-line assembly for surface mount technology requires a uniform PC board shape. That is not to say that all boards must be the same size, because the conveyor systems on the assembly equipment are adjustable. Although adjustable, each manufacturer's system might have a unique maximum or minimum board size limit. When processing the board with an in-line assembly system, specific features and clearances must be recognized. For example, a portion of the board edge is reserved exclusively for conveyor handling. Providing a restricted zone on two parallel edges of the circuit board is a prerequisite for conveyer transferring of the assembly from one machine to another (see Fig. 5-2).

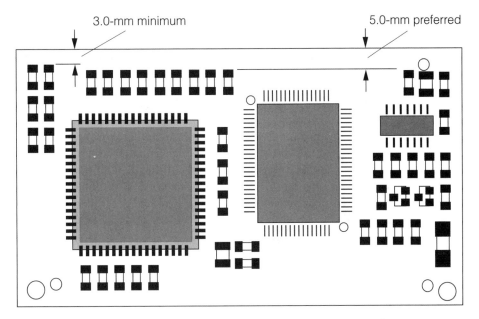

3.0-mm minimum

5.0-mm preferred

■ **5-2** *When planning for in-line assembly processing, the designer must consider the components physical relationship to the edge of the circuit board.*

Panelization for efficient assembly processing

The in-line SMT assembly systems in use today are designed around a substrate size that is generally within the vast majority of products. There is no industry standard for panel size but each assembly system will have a maximum size limit. For example, 355×456 mm (14 in $\times$ 18 in) is the maximum dimension of a substrate processed using the Fuji CP II, but other models might allow 456×507 mm (18 in $\times$ 20 in) or larger panels. The detail shown in Fig. 5-3 will accommodate several assembly systems.

When components must be placed close to the board edge, special fixturing is generally employed to hold the assembly, but this might add cost to the processing of the product. The addition of a temporary tab extension or breakaway tab on one or two edges of the board might be a solution, but this too can add additional cost to each PC board.

Providing for chip device attachment

When using reflow solder processing, each device must have a dedicated land pattern set or array. The attachment of two or more passive devices to a single-shared land pattern, for example,

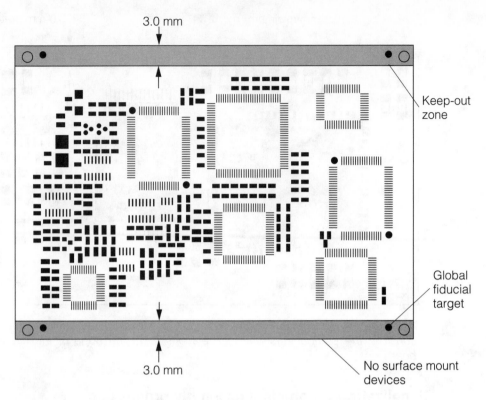

3.0 mm

Keep-out zone

Global fiducial target

No surface mount devices

3.0 mm

■ 5-3 *Although the edge area is reserved as a component keep-out zone, locating global fiducial targets within these zones is often acceptable.*

is not acceptable unless the devices are attached first with epoxy for wave-solder processing. During the reflow process, the volume of solder on the common-shared attachment site will overpower both devices pulling them toward the center of mass as shown in Fig. 5-4.

If devices must be connected in series as the example illustrates, the designer should separate the land patterns with enough space to allow for solder mask and use one or two narrow circuit traces for the interconnect as detailed in Fig. 5-5.

Component spacing

During the reflow-soldering process, the solder transforms into a liquid form. While the solder is liquid, the components float on the high point of the contact pad. Problems can occur when one component is spaced too closely to the next. One component might draw toward the other or slide off the center onto the adjacent

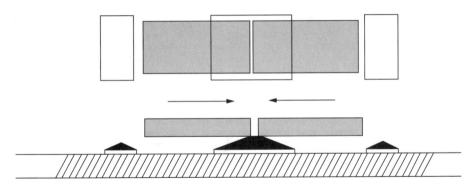

■ **5-4** *During the reflow process, the volume of solder on the common-shared attachment site will overpower both devices pulling them toward the center of mass.*

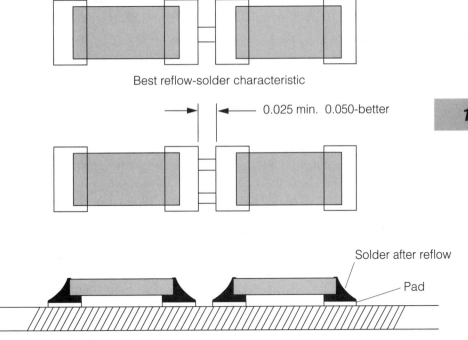

Best reflow-solder characteristic

0.025 min. 0.050-better

Solder after reflow

Pad

■ **5-5** *Spacing between components that are connected end-to-end must provide for solder mask separation of SMT land patterns.*

pad. This problem generally is eliminated with adequate clearance between land patterns, as shown in Fig. 5-6.

To compensate for less than ideal pad geometry, it might be necessary to use component-mounting epoxy normally reserved for

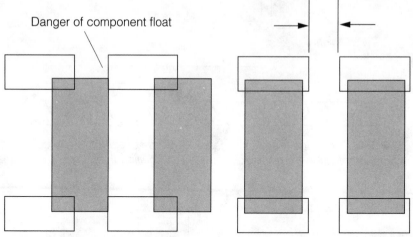

0.65 mm (0.025 in.) min
1.27 mm (0.050 in.) better

Danger of component float

■ **5-6** *Spacing between chip devices, if too close, may allow component shift during assembly. Clearance should be uniform and allow for solder mask separation.*

wave-solder attachment. Adding epoxy to the reflow-solder process as a band-aid for poor design increases the cost of the assembly.

There are two other disadvantages of using epoxy in the reflow procedure. First, particles from the solder or flux can be trapped in the epoxy material. Also excessive epoxy can flow onto the land pattern area and can compromise the quality of the solder joint.

Land pattern to via pad connection

Space between the device land pattern and via hole pad must be wide enough to accommodate solder-mask material. When a land pattern and via pad must be joined without adequate space, it will be necessary to cover the via pad with solder mask. In reflow-solder processing, if the via hole is connected directly to a land pattern and is not covered with solder mask, the liquid solder will flow down through the hole and away from the intended component contact.

Via hole pads against or within the land pattern, if not plugged, also will cause migration of liquid solder away from the contact area during reflow. Via holes and pads under chip components are not recommended because they cannot be seen, as shown in Fig. 5-7, and they can cause failure during additional processes. Specifying a smaller hole size and plugging might be an alternative to

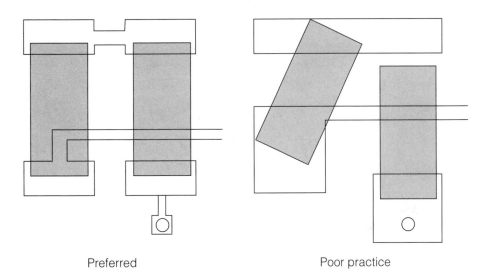

Preferred   Poor practice

■ **5-7** *Chip component land patterns should be interconnected to other features with a narrow circuit, allowing equal surface tension through heat rise and cooling of the solder alloy during the reflow process.*

the separation of contact and via pads, but it might affect the cost of the PC board.

Via holes and pads or heavy circuit traces adjoining the contact area are not recommended for reflow solder; however, these factors will not have negative results in the wave-solder process. A via hole and pad under a chip device without solder-mask tenting, as shown in Fig. 5-8, is not recommended. Solder or adhesives can migrate into the hole during the process, thereby causing a defect that will require additional rework.

Containment of an equal volume of solder on the component land pattern is vital to provide surface tension. When solder volume is not equal at each end of the chip device, separation or de-wetting might occur. The migration of solder when in its liquid state, as detailed in Fig. 5-9, will cause one end of the component to pull away or shift from the contact area.

It is a common practice to cover or tent over closely connected via holes that are not test probe contacts with solder mask. The solder paste will be restricted only to the contact area of the device during the reflow-solder process.

For reflow-solder processing, the width of the land pattern for chip components should be the primary consideration. If the land

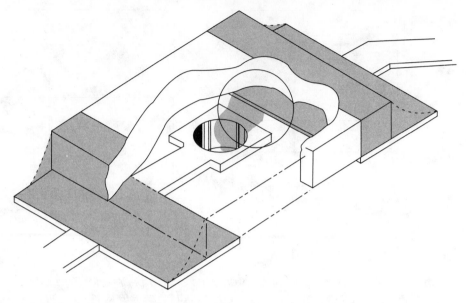

■ **5-8** *Via hole (and pad) under chip resistors and capacitors is not recommended. Relocate the via in an area that is clear of the component body.*

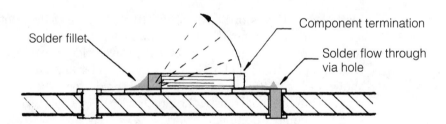

Solder fillet

Component termination

Solder flow through
via hole

■ **5-9** *Solder bridging under a chip component can be avoided by covering a via pad with solder mask or relocating the via to a clear area outside the device's body.*

pattern is too wide, the chip component had a tendency to rotate, as illustrated in Fig. 5-10.

Likewise, a land pattern that is too long will cause the chip component to float off the contact during cooling of the liquid solder material. The ideal pattern allows the liquid solder to have equal surface tension, evenly centering the component on both attachment sites.

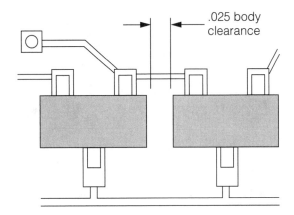

.025 body clearance

■ **5-10** *Space between SOT-23 device bodies must allow for machine placement without physical interference.*

Interconnecting SOT-23 components

Reflow soldering of surface mount transistor (SOT) devices requires a minimum space between pad area contacts as compared in Figs. 5-10,11, and 12. The distance specified in Fig. 5-10 furnishes an adequate solder mask area to prevent solder migration away from the component lead. Maintaining distance between contacts of the SOT device also reduces excess solder buildup, bridging, and voids when the wave-solder process is applied.

Solder mask for solder control

Solder mask, a valuable ally in the reflow assembly of surface mount products, can be applied with either a dry film photo-imaged lamination process or liquid photo-imaged polymer, or a combination of liquid and film. Clearance of a photo-imaged solder mask to land pattern can be zero, but with 0.13–0.25 mm (0.005 in–0.010 in) overall expansion. This clearance can be controlled by furnishing an expanded pad master to the board fabricator with instructions not to expand further.

The two examples shown in Fig. 5-13 compare the ideal solder-mask clearance to the unacceptable. Solder mask options are detailed further in chapter 8.

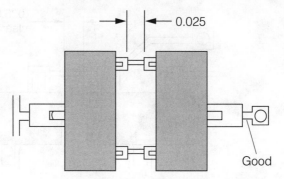

0.025

Good

■ **5-11** *The spacing provided for the SOT-23 land patterns and via pads will ensure solder containment.*

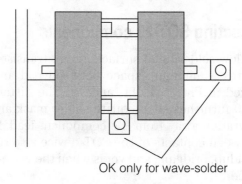

OK only for wave-solder

■ **5-12** *Joining the contact land patterns of adjacent components and vias is not recommended for reflow-solder processing.*

Except for the single, one-sided PC boards, the designer should always specify a photo-imaged mask coating. When the wet screen pattern application of solder mask overlaps the contact area or when solder mask residue is left on the land pattern area, reflow soldering will not be satisfactory. Likewise, when too great of a clearance is allowed, the liquid solder will spread away from the contact area, promoting an unreliable solder connection.

When a via is to be used as a test probe contact use a via pad size of 0.9 or 1.0 mm (0.035 or 0.040 in) round or square, and if space permits provide a minimum center-to-center space of 2.5 mm (0.100 in) between test probe contact locations. See Fig. 5-14.

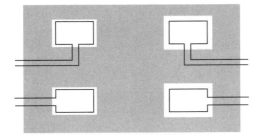

5-13 *Solder mask clearance around the contact must be kept to a minimum. A wide opening will allow solder paste migration when the alloy is in a liquid state.*

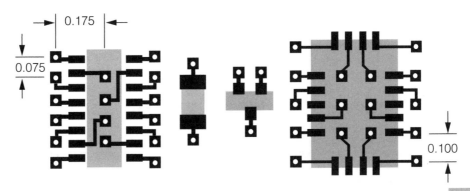

0.175

0.075

0.100

5-14 *Via pad and hole spacing can be optimized for efficient access by automated test systems.*

Do not position via pads under small resistors or capacitors. Migration of solder during the reflow process will cause the liquid metal to draw away from the contact area. Details shown in Fig. 5-15 further illustrate the danger of solder bridging under the body of the chip component. This defect can occur if a small amount of solder paste is lodged under the component or during post wave-solder operations.

Contact patterns that are positioned too close to via pads will cause additional solder defects. The solder paste, when heated to a liquid, flows into the via hole and away from the contact area. It is common to position via holes and pads under the bodies of ICs. However, avoid hidden via holes and pads near unrelated contact patterns (Fig. 5-16). The only way to correct a bridge or short that might occur under the component is to desolder and remove the device.

Illustrated in Fig. 5-17 are potential process defects related to via holes and pads on SOICs. Figure 5-18 compares a good design

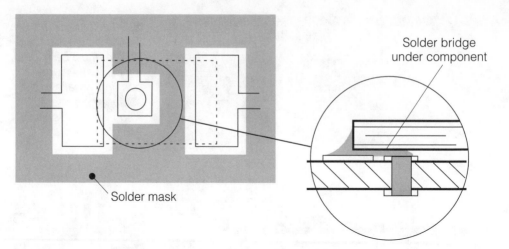

Solder bridge under component

Solder mask

■ **5-15** *Via hole pads placed under the body of the chip component will provide an unwanted conduit for solder migration during wave-solder processing.*

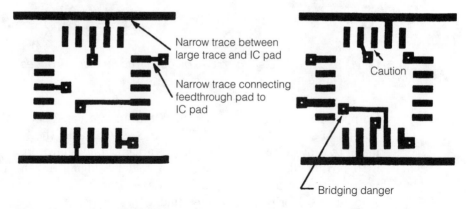

Narrow trace between large trace and IC pad

Narrow trace connecting feedthrough pad to IC pad

Caution

Bridging danger

■ **5-16** *Plated through-holes (or via holes) and pad area must be separated from the land pattern by solder mask in order to contain the solder paste during the reflow-soldering process.*

practice, with clearance provided for solder mask, to a poorer design practice that will require tenting or covering the via holes with solder mask.

Trace-to-trace and trace-to-contact patterns shown in Fig. 5-19 allow for consistent etch quality on outer circuit layers, while ensuring thorough coverage by the solder mask over the circuit trace. An exposed edge of the conductor passing near or between con-

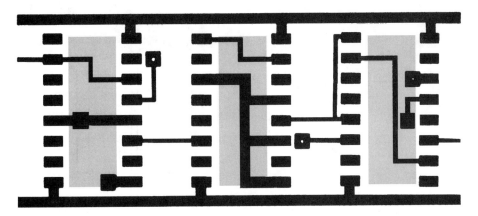

■ **5-17** *Spacing between the via hole pad and device land patterns, if not covered by solder mask, might bridge during assembly processing.*

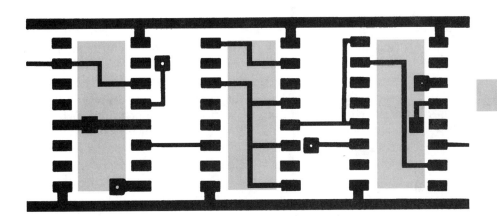

■ **5-18** *By adhering to minimum spacing recommendations on via hole pads, the designer reduces the need for costly rework to remove solder bridging.*

tact areas will attract solder particles and result in bridging. Removal of the component is the only way to eliminate a short when it occurs on a PLCC or SOJ device because the contact folds under the component.

When it is important to increase conductor width between components, the best result is achieved with two narrow traces rather than one wide one. Containment of the solder within the land patterns is the key to controlling the solder process.

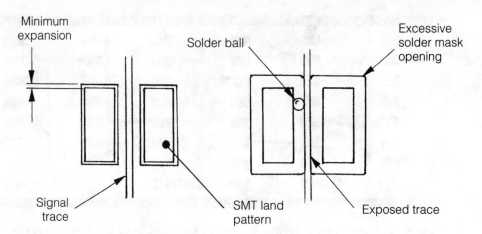

■ 5-19 *Trace-to-trace and trace-to-contact patterns allow for consistent etch quality on outer circuit layers, while ensuring thorough coverage by the solder mask over the circuit trace.*

Tin-lead solder plating of copper traces on circuit boards is not recommended for reflow processing. The alloy under the mask will return to a liquid state during the high-temperature reflow-solder process and, while liquid, the solder will redistribute under the mask coating, resulting in an irregular appearance. When the molten material collects in a mass, cracks or breaks in the solder mask coating will occur. The preferred circuit board for surface mount assembly will have solder mask over bare copper (SMOBC) or over a plating that will not flow at higher temperature, typical of solder processes. The solder plating, if specified, is limited to the exposed contact area of the components, and attachment holes or vias, while conductor traces and other features under the mask material remain undisturbed and flat.

Conductor trace routing

When preparing for photoplotting of closely spaced conductor traces on an SMT board, it is essential to reduce trace width. Routing wide signal lines between the 0.63-mm (0.025-in) wide contact patterns used on SO and PLCC ICs requires reducing the conductor width to 0.20 mm (0.008 in) over-all or, as traces pass between contact patterns, providing an air gap.

Contact (land) to via pad

The space between a via pad or via hole and the component contact area should provide for a solder mask barrier. This barrier will contain the liquid solder paste in the contact area during the reflow process. The configuration shown in Fig. 5-20 provides a 0.20- to 0.25-mm (0.008–to 0.010-in) wide solder mask strip separating the contact from the via pad and hole.

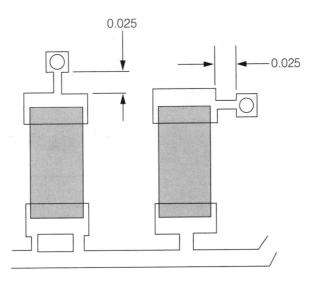

■ **5-20** *Connect the SMT land patterns to a via hole and pad using a narrow circuit trace. Isolation of the contact by solder mask will stop the migration of solder in its liquid state.*

Providing for vision-aided assembly automation

The more advanced assembly systems are equipped with fiducial recognition for visual alignment correction and accurately place surface mount components at high speed within 0.05 mm (0.002 in) of true position. To take advantage of the placement accuracy provided by the automated vision systems on advanced assembly systems, fiducial targets must be added to the PC board artwork. Two or three global fiducial patterns will be required to accommodate automated visual alignment. The examples shown in Fig. 5-21 are accepted by most systems and conform to the recognized Surface Mount Equipment Manufacturers Association (SMEMA) recommendations. The fiducial targets are etched into the panel

during the board fabrication process and will ensure a more precise location tolerance match for each device to the circuit land
pattern.

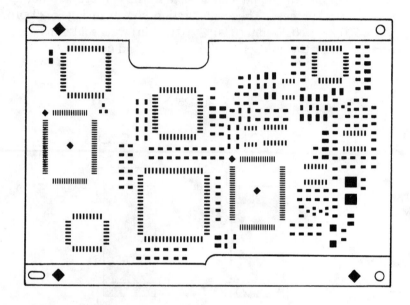

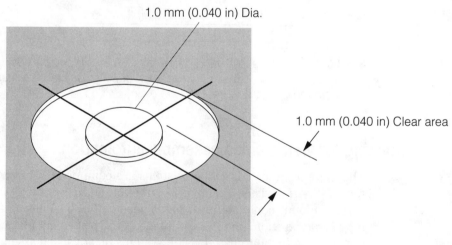

1.0 mm (0.040 in) Dia.

1.0 mm (0.040 in) Clear area

■ **5-21** *Panel construction of one or more units-per-board should allow for machine automation of all assembly processes.*

Mixed technology leaded for through-hole and surface mount

Mixed technology using leaded or pin-in-hole (PIH) with surface mount components is often unavoidable. Until all components are available or economical for surface mounting, the need to continue using leaded-through-hole devices on the same substrate will remain.

Close attention to component spacing is important when the designer mixes PIH parts with SMT. PC board assemblies often require a header or a connector interface to another assembly or segment of the system. The connector products available in a surface mount configuration do not always have multiple sources. Identical SMT connectors from two or more manufacturers are not common, thus forcing the designer to revert to the leaded components. This is also true for the larger value capacitors, resistors, or potentiometers found only in packages with leads. The designer, looking at the layout in one dimension, might fail to consider the component height when positioning surface mount devices near leaded components. As noted, the space between high-profile SMT and PIH parts must accommodate visual inspection and access of rework tools, soldering irons, test probes, etc.

Figure 5-22 details the profile of the PLCC and the typical leaded through-hole device. Touch-up or rework tools require a minimum space to access the solder area without disturbing or damaging the adjacent component.

With all the focus on the implementation of surface mount technology, it is easy to put aside important design rules for the leaded or pin-in-hole (PIH) assembly. The use of automation on the PC board assembly is not limited to the SMT process. The designer should not assume that all PIH devices will be added by hand at some secondary assembly station.

It is the designer's responsibility to make the assembly producible using the most efficient methods possible. Not all assemblies are or will be pure SMT and one should not assume that any component is to be hand loaded, thereby breaking the established requirements of automated assembly systems.

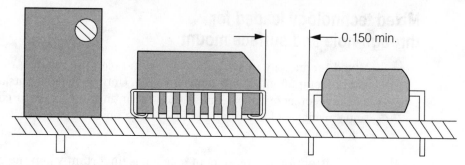

■ 5-22 *The profile of the PLCC and the typical leaded through-hole device.*

Axial lead device mounting

For many electronic companies, axial lead insertion was the first form of automation for circuit-board assembly. It is very efficient, cost effective, and if the circuit board has been designed correctly, it can reduce the need for unnecessary hand labor.

The automatic insertion and crimping of leads on axial devices has been used extensively for decades. In this process, the device is fed to the machine's fixed placement head from a tape-and-reel carrier. The substrate is shifted into the proper X-Y location under the placement head from a zero reference point and the device is inserted, leads crimped, and trimmed.

Hole diameter and spacing requirements for leaded devices

The hole diameter recommended for auto-insertion will allow for various tolerance accumulation. The hole location and size have acceptable tolerance variables. The hole diameter can be ±0.07 mm (0.003 in) after drilling and plating. True center location is allowed to be off by 0.6 mm (0.0025 in). In addition, the tooling holes to locate the circuit board in the machine will have the same tolerance variables. To provide a hole diameter to successfully insert a ¼-watt resistor with a lead diameter of 0.70 mm (0.028 in), the hole diameter in the substrate should be 0.96 mm–1.06 mm (0.038–0.042 in). The spacing between one lead and the other is also critical. The designer must allow for the lead bending and guidance mechanism of the machine. Detail shown in Fig. 5-23 illustrates the basic requirement for lead spacing of the axial device.

Axial lead component placement guidelines

To reduce processing time on the insertion machines, the axial lead devices are often sequentially mounted to a tape-and-reel

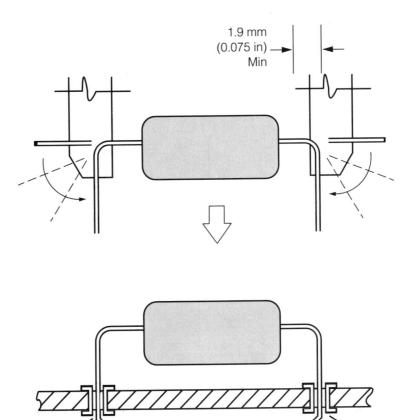

1.9 mm
(0.075 in)
Min

■ **5-23** *Axial leaded PIH devices are suppled in a tape and reel format to allow the lead to be cut, formed, inserted into the holes and clinched to the circuit board.*

carrier. By maintaining a common axis on axial lead devices the designer can make a significant contribution to manufacturing efficiency. Each time that a component changes direction from one axis, more time is added to the machine cycle. Ideally, all axial leaded devices would have a common orientation as those illustrated in Fig. 5-24.

Another practice that will save time and help avoid errors during assembly is common direction or orientation of polarized devices. Components such as diodes, power capacitors, and transistors could, with a concerted effort, be oriented on one axis and one polarity direction.

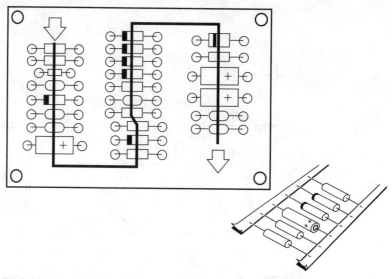

■ **5-24** *Ideally, all axial leaded devices would have a common orientation and polarized devices a common direction.*

Because of critical circuit layout restrictions, some designs might not offer the latitude for the additional circuit trace length to maintain one axis. However, when common orientation is possible, the designer can again significantly improve manufacturing efficiency of the product.

The examples shown in Fig. 5-25 compare typical component layouts. One is more efficient to assemble than the other because of the common axis of parts, even though the polarity is not common on all devices.

Each time the orientation changes on a component, additional machine time is needed to rotate the circuit board or components

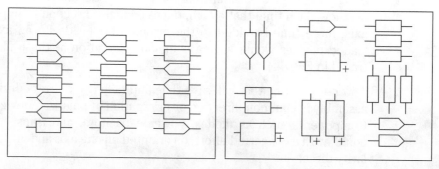

■ **5-25** *Unlike surface mount device attachment, PIH devices with two axis orientation requires the board to be rotated 90°.*

must be specially sequenced. The components for axis "X" must be taped in order of insertion then sequenced again for use on axis "Y".

Axial lead component spacing

The space separating devices on the substrate must be adequate for machine insertion without damage. It's obvious that the outside body must not overlap the adjacent device. The minimum space separating the leads will be significant also. The axial leads are guided through the holes in the substrate, meeting the cut-and-clinch mechanism on the opposite side. The spacing detailed in Fig. 5-26 is the minimum that must be provided to perform cut and clinch without damage to adjacent component leads.

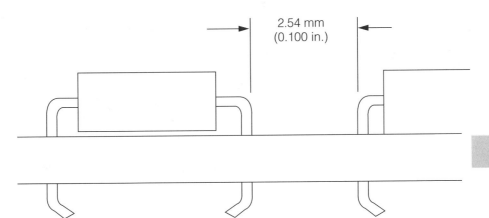

2.54 mm
(0.100 in.)

■ **5-26** *Spacing between lead centers must be maintained to accommodate the cut and clinch mechanism typical of automated insertion systems.*

The machines can be adjusted to crimp leads at almost any angle or against the board. The crimp angle of 45 degrees is generally preferred but the finished clinch length will vary due to the components wire diameter.

If surface mount devices are to be attached for wave solder on the clinch side, excessive lead length might cause bridging, as shown in Fig. 5-27.

Design guidelines for DIP ICs

The use of DIP ICs on surface mount assemblies has been dramatically reduced with the SO (small outline) IC. Most of the common logic and analog functions are available in the miniature gull-wing package and are at price parity with the DIP device they replace.

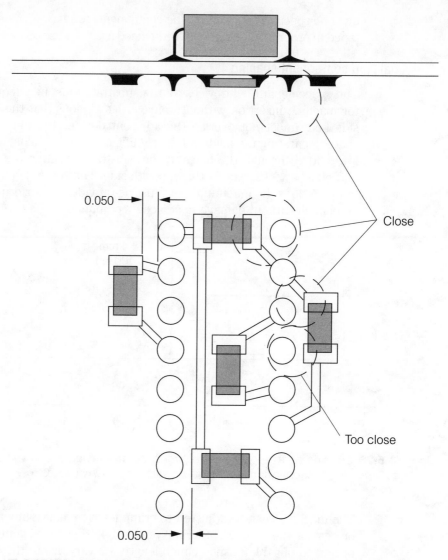

0.050

Close

Too close

0.050

■ **5-27** *When attaching PIH devices for wave solder, maximize the spacing bewteen the protruding device lead and surface mount device body to avoid solder bridging.*

Some devices, however, have not been made available in surface mount. An example is the ECL family and the 74S series of ICs. In other cases, DIP packages are retained by choice. It is still common to socket EPROMs and other devices. The standard lead type DIP socket remains a bargain when compared to the alternative surface mount DIP socket.

When several DIP ICs are designed into the assembly, common orientation of these components is preferred. As with axial leaded devices, when the orientation changes, the board must be rotated on the machine before the assembly can continue.

The detail shown in Fig. 5-28 is typical of the insertion order on a circuit board. Because of space restraints or circuit layout restrictions, some of the DIP devices must rotate. However, it is considered poor design practice to orient DIPs in more than two different directions. When three or four IC directions occur, strong justification is generally offered to defend the practice ("not enough time" is not justification).

Hole size recommendation for DIP ICs

The hole size recommended for IC leads is nominally 1.0-mm (0.040-in) diameter. The hole diameter must allow for tolerance variables in both the substrate and the machine used for assembly. The designer must provide an adequate annular ring around the finished hole to maximize PC board fabrication yield because breakout of holes on pads on the external surface is not acceptable. When sockets are designed into the assembly, use additional care during selection. Some sockets, especially those that press directly into the substrate, require a finished hole diameter far greater than the ICs.

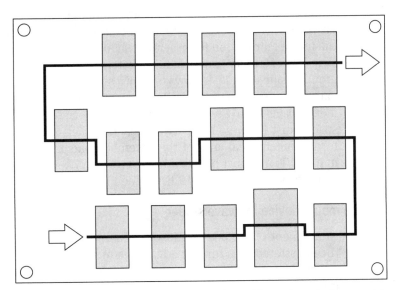

■ **5-28** Uniform IC placement is preferred for automated assembly efficiency while a nonuniform multiple axis, IC placement will increase assembly process time.

Spacing between DIP ICs and sockets

Spacing between rows of DIP ICs can be as close as 2.5 mm (0.100 in) if sockets are not required. If space permits, allow more clearance. Additional space will accommodate test probes and the use of sockets during the development test phase of the program. Figure 5-29 illustrates the minimum spacing required for the cut-and-clinch mechanism on the insertion system.

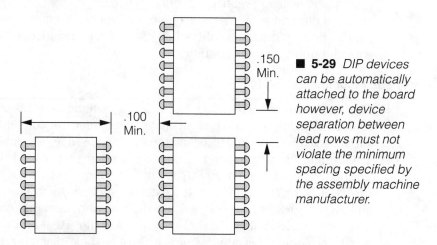

■ **5-29** *DIP devices can be automatically attached to the board however, device separation between lead rows must not violate the minimum spacing specified by the assembly machine manufacturer.*

DIP socket spacing for auto insertion

When sockets are required for programmable devices, upgrade provisions, or memory expansion of computers or other products, the use of automation for assembly should be planned into the design. The memory expansion sockets, for example, are usually grouped side by side, in rows of 8 or 9 sockets each. To accommodate auto insertion of the DIP sockets, the designer must provide more space than the DIP devices without a socket. Additional space must be provided between each socket to allow clearance for the mechanical jaws of the assembly machine. Spacing of 2.5 mm (0.100 in) must be provided between socket bodies (not lead centers as specified for DIP ICs). See Fig. 5-30.

Surface mount device for wave solder

Attachment of surface mount devices on the wave solder side of the substrate takes further advantage of surface area. The epoxy attachment of these devices usually follows all insertion and crimping operations.

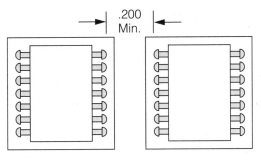

.200
Min.

Socket mounted
integrated circuits

■ **5-30** *DIP socket location must allow additional spacing for the gripping of the insertion mechanism on both sides of the socket body.*

Surface mount bypass capacitors and other passive devices are often positioned to use the space between lead rows of sockets. This common technique will take advantage of surface area not needed for circuit routing.

The illustration shown in Fig. 5-31 is typical of the techniques used to incorporate the low-profile surface mount devices on the wave solder side.

Clearance must be provided to avoid solder bridging between the leads protruding through the board and surface mount contact area. Solder mask separation is generally adequate, but when leads protrude too far and are bent toward the surface mount component, solder bridging can seldom be avoided.

Chip component land pattern options for wave solder

A land pattern that is too large for the chip component will encourage an excess of solder buildup. Excessive solder on the chip device can cause physical stress during thermal cycling, resulting in cracking the device. When wave soldering discrete components, it is possible to adapt an optional narrow pad geometry to limit the amount of solder volume at the component connection.

Be aware of the component body width when positioning the land pads. If the narrow pattern is used, it will appear that the designer has adequate clearance for routing traces or via placement, but

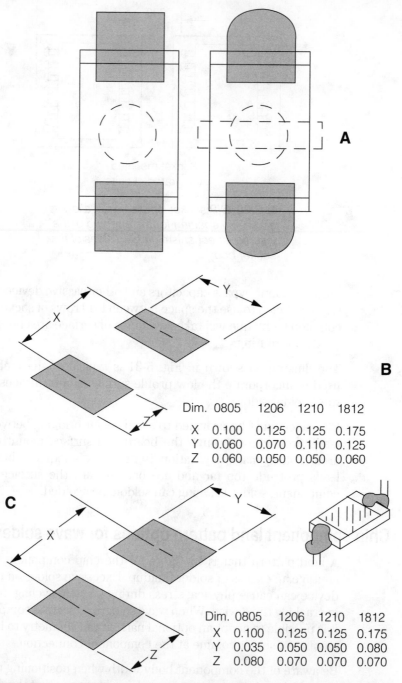

Dim.	0805	1206	1210	1812
X	0.100	0.125	0.125	0.175
Y	0.060	0.070	0.110	0.125
Z	0.060	0.050	0.050	0.060

Dim.	0805	1206	1210	1812
X	0.100	0.125	0.125	0.175
Y	0.035	0.050	0.050	0.080
Z	0.080	0.070	0.070	0.070

■ **5-31** *(A and B) Two examples of passive device land pattern variations specifally designed for wave solder attachment. (C) the optional narrow land pattern for chip resistors and capacitors.*

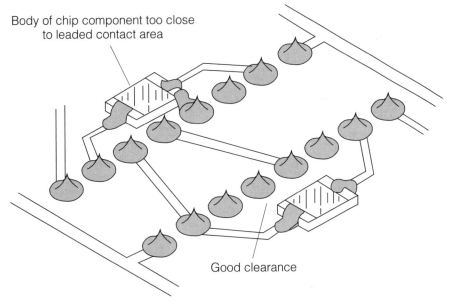

Body of chip component too close
to leaded contact area

Good clearance

■ **5-32** *Solder bridging is common when inadequate spacing is provided between epoxy attached surface mount device and the lead of devices mounted from the opposite side.*

when a component is placed, the potential for solder bridging increases (Fig. 5-32).

Device selection guidelines for wave solder

Passive low-profile devices are generally preferred for adhesive attachment and wave-solder processing, but it's not uncommon to attach small outline ICs on the secondary side as well. Because of concerns related to exposure of the plastic housing to molten solder, some companies might not allow active device attachment for wave solder due to the increased component density on some PC boards. However, it might be necessary to make a decision to either create a piggy-back assembly or transfer-selected ICs to the secondary side of the board. Although reflow soldering of both primary and secondary sides of high-density surface mount assembly is a common practice, wave soldering is more economical, but component spacing and orientation must be well planned.

Component-to-board edge
requirement for mixed technology

PIH devices can be mounted close to the substrate edge, but as with the reflow soldered assembly, the designer should avoid close edge placement, especially when the assembly is to pass through a wave-solder process. When wave soldering is to be part of the assembly sequence, and adequate edge clearance cannot be furnished, special pallet fixtures can be developed or break-away strips can be added to the long edges of the board to provide an ample holding surface for the machine's conveyer mechanism as shown in Fig. 5-33.

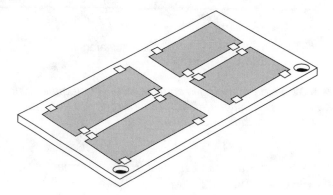

■ **5-33** *Individual substrate units without compatible tooling holes must be mounted to a customized fixture designed to adapt to a specific assembly system.*

Figure 5-34 illustrates how component body to substrate edge clearance must allow for direct machine handling. For those assemblies with a restricted edge clearance, a break-away strip can be added to each unit.

Alternative reflow/wave solder processes

When the component density of the product mandates the use of both sides of the board for component attachment, two reflow processes might provide the best control of solder defects. It is recommended that fine-pitch (QFP, SQFP, and TSOP) devices be confined to only one side. Often due to space limitations or the functional relationship between specific devices, the parts are

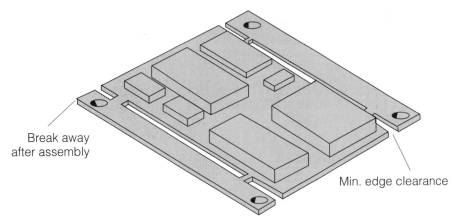

Break away
after assembly

Min. edge clearance

■ **5-34** *If components are mounted close to the board edge preventing conveyor transfer, temporary extensions of the PCB material can added.*

mounted to both primary and secondary sides. Although two-sided assembly will require exposing the board to the reflow process twice, the defect ratio is minimal.

If the same high-density, two-sided surface mount assembly has several leaded PIH devices, connectors for example, it might require a third soldering operation. Hand solder is an option, but the cost associated with hand soldering can be prohibitive, especially in high-volume assembly. The manufacturer might prefer to use an automated solder process for attaching the leaded parts. Robotic systems are available for programming soldering at each site, but the process can be too slow if the lead count is high.

Other choices include *solder pot* and *solder fountain* systems, but both require excessive handling. The wave solder continues to be the most economical method of mass termination of leaded devices mounted after reflow solder. This technique of using double reflow followed by wave soldering will require some unique fixturing and, most important, planning at the board design level.

Planning for double reflow-wave

The designer must provide additional spacing between the leaded device and surface mount devices attached to the secondary side. The space is needed to allow for fabrication requirements of the

133

wave-solder fixture. The fixture will act as a mask, covering the already soldered surface mount devices. Openings will be cut into the mask for lead exposure to the wave; a small web of material remains preventing the wave from flooding on the board surface.

SMT layout and guidelines for flexible circuits

FLEXIBLE CIRCUITS ARE UNIQUE IN BOTH SHAPE AND application having specific flex-cycle requirements depending on product application. Before developing the flex circuit the designer must know and understand fabrication methods, material limitations, and assembly processes. By replacing wire and cable (usually the most unreliable element in the interface of active circuit assemblies), flex circuit assemblies furnish a tough reliable interconnection. By incorporating surface mount components onto the flexible substrate, the interface vehicle will become the primary active circuit. Connector and discrete cable interconnection is reduced or eliminated with flexible circuit assemblies. By adhering to recommended design guidelines and early interaction with the circuit fabricator, the user will yield measurable benefits. Benefit to the product manufacturer will be furnished in both reduced material and assembly labor cost. Benefit to the end user will be in the rugged, long-term reliability of the finished product.

Some circuits, for example, can be formed into a fixed position with no on-going flexibility requirement. In complex mechanisms, flexibility and life cycles of the flex cables must be engineered into the design. Three typical product categories for flexible-circuit assemblies with radically different flex-cycle requirements are represented by the following:

☐ Camera products   Less than 5,000 cycles

☐ Printer head   20 million cycles

☐ Disk drive head   400 million cycles

Rigid materials are often integrated with flexible material to provide a firm base for attachment of surface mount devices, although attachment of devices directly to an unsupported flexible base is also common.

Base material selection

Issues to be considered when planning the flexible-circuit assembly is base material options and cost of fabrication. Because the fabrication process for flexible circuits is somewhat specialized, unique design rules will apply when adapting this medium. Several materials are available for the fabrication of flexible circuits, but selection will be dependent on the end use environment and assembly processes to be incorporated.

Base material properties for flexible circuits

☐ *Polyester* is low cost and has excellent electrical property and flexibility, but it has poor resistance to temperature exposure typical of SMT solder processes.

☐ *Polyimides* are high in cost, but have excellent dielectric and physical strength and flexibility; they are compatible with solder process temperatures of up to 230° C, but material can absorb moisture (up to 3% by weight).

☐ *Fluorocarbons* are less costly than polyimide, with excellent flexibility and good dielectric; they have low moisture absorption but soldering process temperatures must be less than 180° C.

These materials are available with several conductive metals for circuit applications. Conductor alloy choices include aluminum, copper, gold, silver, and nickel. Copper alloy is typically the choice for etched circuit fabrication because of its compatibility with rigid board fabrication, superior conductive properties, and adaptability to solder attachment processes.

Preferred materials for SMT solder processing

The most common base material used for SMT applications is 0.05-mm to 0.07-mm (0.002-in to 0.003-in) thick polyimide with copper foil laminated on one or two surfaces. The copper foil thickness is usually 0.018 mm (0.0007 in) ½ oz. to 0.035 mm (0.0014 in) 1 oz.; however, 0.008 mm (0.0003 in) ¼ oz. copper might be specified for very fine circuit features. After the circuit pattern is etched to remove excess copper, an insulating layer or cover layer is laminated over the copper pattern. The example shown in Fig. 6-1 is typical of the layered material, only the SMT land pattern and other solder contact areas remain exposed.

Although alternative, less costly, materials to polyimide are noted for flexible-circuit applications, they are not compatible with the

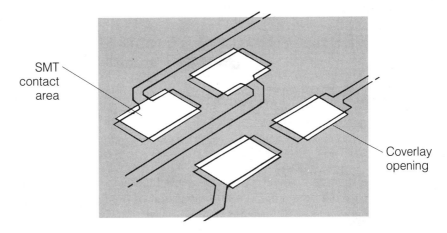

SMT contact area

Coverlay opening

■ **6-1** *SMD land pattern geometry must be extended for overlay entrapment. This will ensure soldering and resoldering of a component will not delaminate the copper from its base material.*

extreme temperatures of solder assembly processing typically employed for surface mount technology.

Fabrication of flex and rigid-flex circuits

The fabrication procedure for laminating and etching the flexible and rigid-flex substrate circuit is similar to methods utilized to process multilayer rigid substrates, but materials used to fabricate the flexible circuit are quite unique. Surface mount technology requires a dielectric material that can withstand the 200–230° C temperatures experienced during solder processes. Polyimide films are favored because they meet the demands of assembly processing and remain tough, easily conforming to any physical shape or environment.

Rigid support for flex circuits

To improve physical characteristics of the flexible circuits, a rigid backing material can be laminated in selected areas to support heavy components or to provide mechanical stiffening. The flex circuit also can be integrated into a rigid circuit board to furnish a flexible interface between connectors or other areas requiring a rigid substrate. In regard to flexible circuits requiring a rigid backing, the flex circuit can be bonded directly to the preformed rigid material. Adhesive films are precut to match the rigid section and pressure laminated to the specified area requiring the stiffener. See Fig. 6-2.

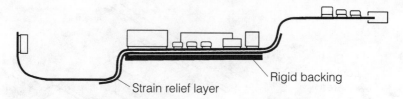

Strain relief layer

Rigid backing

■ **6-2** *Adhesive films are precut to match the rigid section and pressure laminated to the specified area requiring the stiffener.*

If the unsupported extension must flex near the edge of the rigid base material, consider adding a layer of the flexible base material between the circuit and rigid material, as shown, to function as a strain relief and tear restraint.

Rigid-flex circuit construction

When the rigid portion of the substrate includes circuit layers, the interface between the rigid and flexible material should be designed as a balanced structure. Ideally the etched flexible portion of the composite is laminated or sandwiched between equal thicknesses of rigid layers. The rigid material can be one or more copper layers and must be fabricated from materials compatible with the base material. The final lamination of the flex circuit to the rigid circuit section is similar to the methods used to fabricate multilayer PC boards with final drilling and copper plating of holes to complete the interface of circuit layers as shown in Fig. 6-3. Via holes, related only to flexible layers of the rigid flex board, are processed before final lamination. Common via holes to both the rigid and flexible layers are drilled and plated after lamination.

Via holes related only to flexible layers of the rigid-flex board are processed before final lamination.

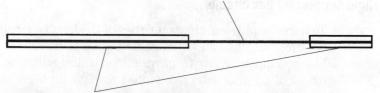

Common via holes to both the rigid and flexible layers are drilled and plated after lamination.

■ **6-3** *The final lamination of the flex circuit to the rigid circuit section is similar to the methods used to fabricate multilayer PC boards with final drilling and copper plating of holes to complete the interface of circuit layers.*

To provide a reliable product and maximize yield in the manufacturing of the circuit, a balanced or symmetrical construction is recommended for multilayer Rigid-Flex printed circuits. As with one- and two-layer flex circuits, a cover layer of the polyimide base material is laminated over the copper center sections to protect the flexible section while conventional photo-imaged solder mask is typically applied to the rigid sections of the circuit.

Flex circuit design guidelines

The design guidelines for SMT on flexible circuits are, in many ways, the same as those used in rigid printed circuit board applications, but some variations are necessary to accommodate the flexible circuit construction. The following addresses issues that can affect the durability and reliability of the flexible circuit product:

☐ Conductor trace routing and bend radii conventions.

☐ Fold line considerations.

☐ Trace connections and hole/pad filleting.

☐ SMT land pattern guidelines and cover layer openings.

☐ Tear restraints and strain relief methods.

☐ Panelization and fiducial targets for SMT automation.

Conductor trace routing

Single-sided flexible circuits have only a few limitations as far as conductor width and spacing. The notably critical factor is in the path of the circuit trace. A change in direction for a conductor ideally will have a gentle radius or 45° angle at the point of the turn as shown in Fig. 6-4. Right angle turns of circuit traces must be chamfered or provide a radii to prevent separation or tearing of the base material.

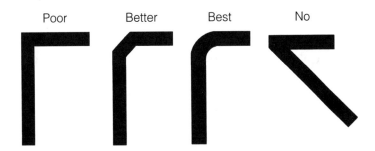

Poor   Better   Best   No

■ **6-4** *Sharp corners of a conductor trace on a flexible substrate will cause tearing of the base material.*

Fold line considerations

Parallel trace runs should always be perpendicular to a fold line. Bending the copper on the bias is a poor practice that adds stress to the trace and can damage or crack the circuit conductors, as shown in Fig. 6-5.

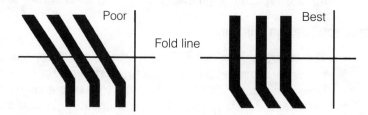

■ **6-5** *If a bend in the base material is necessary, conductor paths should be perpendicular to the fold line. Folding the circuit traces across a bias direction should be avoided.*

The above illustration compares fold line to circuit conductor path. Folding or sharp bends on flexible circuits should be minimized due to the stress of deforming the copper foil, but when required, avoid an angular fold across the conductor path.

Two-sided conductor routing

For maximum flexibility on flex circuits having two circuit layers, parallel trace runs (depicted in Fig. 6-6) are offset or staggered from one side to the other. If conductors are directly opposite one another on a fold line, a stress is transferred to the trace as it attempts to stretch or compress around the bend.

Circuit conductor width and air gap

Minimum conductor width/air gap between conductors can be as narrow as 0.05 mm (0.002 in) on 0.018-mm(0.0007-in) thick copper foil. However, a wider trace and air gap dimension is desirable. A conductor/air gap width of 0.2 mm (0.008 in) or wider will give the fabricator a far better yield on the finished flex circuit.

Preferred conductor width and space

Copper material	Photo imaged pattern
½ oz., single sided, nonplated	0.13/0.20 mm (0.005–0.008 in)
1 oz., single sided, nonplated	0.15/0.25 mm (0.006–0.010 in)
1 oz., single sided, plated	0.20/0.30 mm (0.008–0.012 in)
2 oz., single sided, nonplated	0.20/0.30 mm (0.008–0.012 in)
double sided with plated holes	0.25/0.30 mm (0.010–0.012 in)

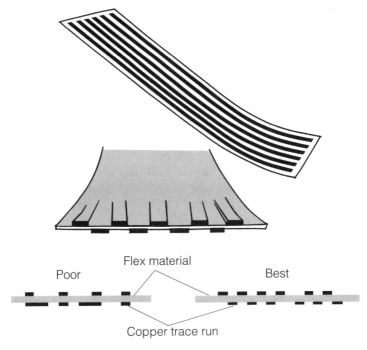

Poor | Flex material | Best

Copper trace run

■ **6-6** *If a fold line is required, conductor trace paths on two sides of the base material should be staggered from one side to the other.*

Copper foil thicknesses, greater than those noted, are available for high current applications, but the thinner foils will furnish the most flexible circuit and enable finer finished conductor lines and spaces. If a two-sided flex circuit is required, the copper thickness specified should be the same on both sides of the base material.

Hole size to pad recommendations

Holes are either punched or drilled in the flexible material and because the overall dimensional tolerance is usually greater than rigid materials, a larger annular ring or copper material around the finished hole size must be provided. Unsupported holes should allow for cover layer entrapment. The pad diameter recommended for plated through holes is 0.76 mm (0.030 in) larger than the hole, furnishing a 0.38 mm (0.015 in) annular ring (copper retained around specified finished hole diameter) and the cover layer opening diameter will be 0.38 mm (0.015 in) larger than the pad.

Because the shape and size of the flexible circuit can vary a great deal, the designer should confirm the final requirement for annular ring size with the fabricator selected to furnish the circuit.

Trace connections and filleting

When narrow conductors meet the attachment feature, the conductor trace should taper or widen as it joins the larger SMT land pattern or mounting hole pad. The widening of the conductor adds physical strength at the interface junction to the larger shape. When joining trace and via hole on a flexible circuit, the annular ring around the finished hole diameter might be smaller than recommended. However, 0.25 mm (0.010 in) or greater ring is preferred. Nonsupported pads or copper areas designated for solder attachment should be entrapped with the cover layer material. For example, fillet junctions can be tapered to large pad or other features added to the smaller pad and hole as demonstrated in Fig. 6-7.

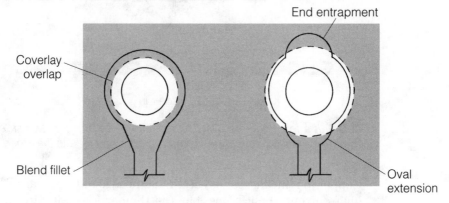

■ **6-7** *The copper area around a nonplated hole will separate easily from the base material if not entrapped by the overlay material.*

One shape, popular with designers and fabricators, is the oval extension on the via pad. The oval pattern can be added to any contact area providing greater bonding strength when contained under the cover layer material. When space permits, simply extend or enlarge the contact area to allow for a perimeter, or end entrapment, of the copper material under the cover layer. This overlapping reduces the occurrence of separation of the copper foil when heated and re-heated by soldering tools or other processes.

Cover layer openings for SMT land patterns

The copper land pattern geometry for SMT will be slightly different from the contact pattern used on conventional rigid board etched circuits. The PC board will generally have an epoxy or polymer solder mask coating applied with open features to expose the SMT footprint. To insulate the conductor traces of a flexible circuit, the cover layer material is laminated to the flexible base circuit with open features prepunched or drilled, as shown in Fig. 6-8 exposing the SMT contact patterns.

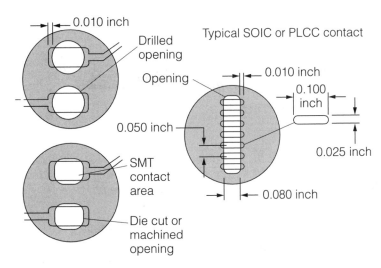

■ **6-8** *Openings in coverlay materials can be drilled, machined, or die-cut to expose the SMT land patterns. In each configuration, a 0.010-in coverlay overlap of the land pattern is provided to strengthen the solder area.*

Designers must modify all extensions to the contact geometry as required to entrap the copper material under the cover layer for strength exposing only the standard SMT land pattern. The cover layer should overlap the footprint by a minimum of 0.25 mm (0.010 in) at each end, but greater overlap is preferred when space permits.

Strain relief and tear restraint methods

If the flexible circuit is going to be exercised excessively at one or more points, a strain relief or stiffener might be required. For example, an additional layer of the base material laminated to the stress-prone area will strengthen a connector. The thickness of the additional material might be 0.25 mm (0.010 in) if flexibility must be maintained, or rigid materials can be added to specific areas, typical of Fig. 6-9, to maximize the physical strength.

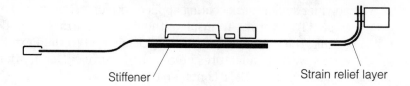

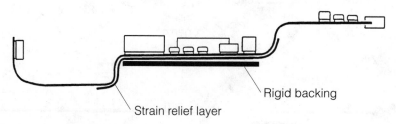

Stiffener

Strain relief layer

Strain relief layer

Rigid backing

■ **6-9** *The thickness of the additional material might be 0.25 mm (0.010 in) if flexibility must be maintained, or rigid materials can be added to specific areas to maximize the physical strength.*

With excessive flexing, the small rectangular SMT land patterns will sometimes promote a tear in the base material after solder attachment of the SMT devices. This damage is caused by the stiffness of the soldered contacts and the weight of components during handling of the unsupported assembly. The addition of a base material layer or rigid backing to the flex circuit in the SMT component-mounting area will provide a more durable and rugged product.

SMT components near a fold line

To reduce the tear factor, avoid mounting SMT components near a fold line. A clearance of 2.5 mm (0.100 in) or greater from the component land pattern to the fold line will reduce the stress between the base material and the copper foil. An extended layer of base material between the rigid-backed portion of the SMT component area to the nonbacked flex circuit will also prevent damage at a fold line. The fabricator should be consulted during the design phase of the product to recommend locations of tear restraints and strain relief layers.

Inside corner support

Tearing in the inside corners of a right-angle bend in the flexible circuit is a common occurrence. To reduce the tearing, provide a generous radius where the circuit shape significantly changes direction and retain a small pattern of the copper foil under the

144

cover layer. This pattern, located inside the finished edge of the part, follows the radii supporting the inside corners. If the direction change requires a thin cut, add a hole at the end of the cut to prevent end tearing, as illustrated in Fig. 6-10.

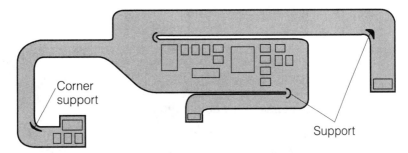

Corner support

Support

■ **6-10** *Retaining small patterns of the etched copper at corner locations will furnish a barrier against tear points.*

Panelization for automated assembly processing

When a flexible circuit is fabricated several individual circuit images are usually repeated as an array by the circuit manufacturer. The final shape or profile for each circuit is exercised from the array by a hard die or steel-rule die as a final procedure.

Small flex circuit assemblies with SMT devices attached using solder technology is difficult to process as a single unit. Even when using special holding fixtures, the flexible circuit material must be secured flat for solder paste application, component placement, and solder-reflow processing.

Automated assembly systems developed for surface mount assembly are designed to handle rigid panels and if flexible circuits are retained on a low-cost rigid backing and assembled before the final excising function, assembly cost and efficiency will be improved significantly. The rigid panel-backed format, all tooling holes, and other features can be included to improve solder paste application, device placement, mass reflow-solder processing, cleaning, and inspection. Although the above described panel implies several small flex circuits retained on a single backing, a more complex or large flex circuit also can be retained on a rigid backing for efficient assembly processing.

A typical configuration would retain partially blanked units in a rigid panel format with small tab connections interlocking individual circuit units. A precut adhesive pattern retains the flexible sheet to the rigid backing only in the area outside the finished part. The profile die would pierce and blank the flexible base material in the finished shape except for a number of tab retaining points, leaving the backing relatively undisturbed. The tab supports are cut or punched after all assembly processes are completed and the rigid backing is discarded. Other materials are available that will retain the adhesive on the rigid backing while releasing the finished assembly. The designer must work closely with both the supplier of the circuit and assembly-process engineers to evaluate and test materials to ensure they meet all process conditions.

Fiducial targets for assembly automation

Vision systems for in-process handling of SMT assemblies have become commonplace in the automated factory. High-resolution video cameras are integrated with the manufacturing systems to improve speed and accuracy. The camera can scan an entire assembly or specific areas or patterns. Equipment manufacturers have recommended uniform shapes for the optical-fiducial target. As with rigid circuit boards, the fiducial pattern acceptable for most vision-equipped systems is a 1.0 to 1.5 mm (0.040–0.060 in) diameter solid pattern etched into the copper material. This image must remain clear of other conductor trace images and cover layer coating. The clearance around the target should not be less than 0.5 to 1.0 mm (0.060 in) to ensure recognition by the vision system. Figure 6-11 represents a typical panel suitable for in-line assembly processing and the global-fiducial pattern recommended for processing the assembly in an automated environment.

Flexible circuit assemblies and surface mount technology can provide for an almost unlimited product configuration. The materials will conform to the most restricted fixed positions or allow extension for interfacing with animated mechanisms. To ensure the product delivered meets the requirements for the specific application, the designer must furnish complete and detailed documentation and instructions to the fabricator.

Documentation requirements

☐ Material description
☐ Mechanical outline detail
☐ Critical dimensions and tolerances
☐ Construction detail
☐ Panelization requirements

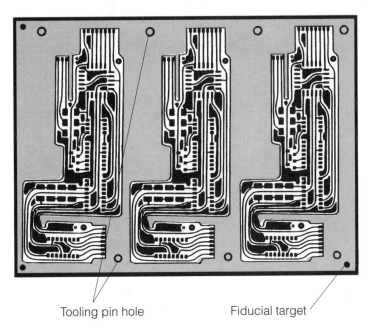

Tooling pin hole   Fiducial target

■ **6-11** *Retaining the flexible circuits in a panel format, with fiducial targets, will assist in each phase of the automated assembly process.*

☐ Gerber data file

☐ Photo tool master

☐ Specifications to be applied (IPC-FC-250)

☐ Define test methods

Although rigid circuits hold a dominant position in the interconnection of electronic assemblies, flexible circuit interconnect systems offer solutions to the very complex and irregular packaging situations. The advantage of the flexible material is its ability to conform to almost any shape. When designed properly, flex circuits have proven to be a rugged, reliable, cost-effective alternative to wire or cable interconnection.

Design requirements for fine-pitch devices

Factors that influence the use of fine pitch

IN ORDER TO CONTROL OVERALL AREA REQUIREMENTS FOR the higher pin-count devices, manufacturers have reduced the lead pitch (the space between lead centers). The standard family of the SQFP has been adopted by the industry to furnish commercial packaging of custom and semicustom integrated circuits with 0.65-mm, 0.5-mm, 0.4-mm, and 0.3-mm lead pitch. Although assembly processing has been refined to acceptable yield levels for 0.65-mm and 0.5-mm pitch device types, conventional process methods might not be practical for high pin count using the 0.4-mm and 0.3-mm pitch devices.

While adapting this advanced packaging technology, the users of the SQFP are challenged by several considerations, including physical, financial, and environmental issues. Physical issues include attachment processes and finished product reliability. Financial decision must be made regarding how the product will be manufactured and the technical level of assembly equipment to be used.

With the growing concerns of environmentally harmful chemicals used in the electronic assembly processes, alternative materials must be considered and tested before commitment to high-volume manufacturing.

Planning the substrate

Many factors impact assembly yield as well as solder joint quality. Each of the following elements must be carefully reviewed and calculated:

☐ Formulate land pattern array and calculate geometry.

☐ Implement the test model substrate design layout.

☐ Detail fabrication specifications and test features.

☐ Define attachment material and assembly process methods.

The small SQFP devices are manufactured in high volume and provide rugged, long-term reliability. Newer packages in the industry, such as the 0.4-mm (0.016-in) lead pitch SQFP might pose a challenge to manufacturers in maintaining a consistent assembly process yield.

Typical device package description

The device package shown in Fig. 7-1 is an EIAJ standard SQFP. The standard 256-pin configuration on 0.4-mm (0.016-in) pitch proves adequate termination channels for more advanced ASIC devices while requiring a relatively small area for attachment.

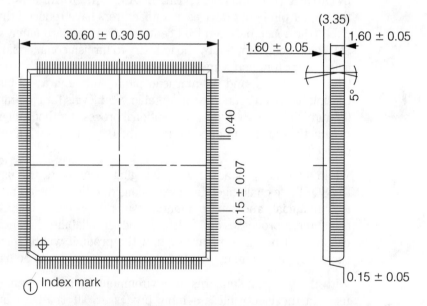

■ **7-1** *The EIAJ, 0.4-mm pitch, 256-pin Quad Flat Pack Device.*

The total surface area (lead end to end) of the 256-lead plastic device is less than 31.0 mm (1.22 in) square with a maximum height from the substrate surface of 3.5 mm (0.137 in). The leads are formed in the gull-wing configuration extending from the device body furnishing a contact or attachment area of 0.5 mm ± 0.2 mm (0.020 in ± 0.008 in). For a process development or test program, a nonfunctional component can be ordered with internal bonding on lead pairs before molding.

Allowance for physical tolerances of devices

Although the lead-pitch tolerance was defined as nonaccumulative, the width of the lead at the attachment area is specified as wide as ±0.07 mm (0.003 in). Using the maximum material condition of the basic lead width of 0.15 mm opens the possibility for up to 0.22 mm total lead width. The final land pattern geometry furnished on the substrate must accommodate this maximum material condition of the device to avoid the possibility of lead overlap at either the toe or heel, as shown in Fig. 7-2.

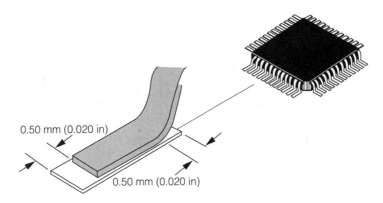

0.50 mm (0.020 in)

0.50 mm (0.020 in)

■ **7-2** *The final land pattern geometry furnished on the substrate must accommodate this maximum material condition of the device to avoid the possibility of lead overlap at either the toe or heel.*

Land pattern development

The land pattern geometry and spacing between pad rows shown in Fig. 7-3 were derived from calculating both maximum and minimum material conditions of the device. That is, if the device is furnished at the maximum overall width (lead end to lead end), the land pattern should provide enough surface area to prevent lead overhang.

Design impact on assembly efficiency

Seventy percent of the failures detected on surface mount assemblies are due to solder defects. Solder defects found at test are ei-

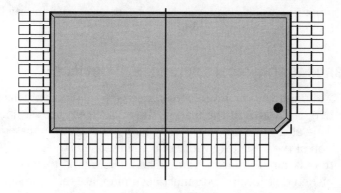

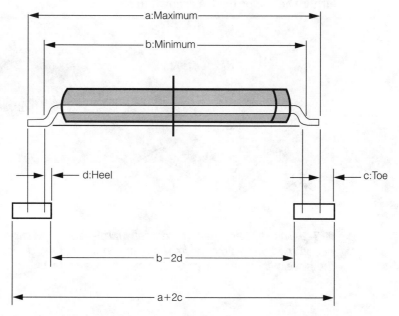

■ **7-3** *The land pattern geometry and spacing between pad rows were derived from calculating both maximum and minimum material conditions of the device.*

ther in the form of a short between device leads or open circuits due to insufficient solder. An ongoing process audit can reduce most of the solder defects, but often manufacturing problems are design related and if not corrected will remain a source of chronic failure.

The most successful SMT programs are those that have implemented the proven circuit board design rules. These rules specifically employ process-compatible land pattern geometry for surface mount devices and each will play a significant role in ensuring manufacturing efficiency and quality. The design rules also will focus on board fabrication guidelines and assembly machine compatibility. Land pattern geometry on the other hand is directly related to process control and solder attachment uniformity. When each of these primary disciplines are properly defined and implemented all solder defect issues can be eliminated.

Avoiding solder process defects

Solder defects can be further reduced through continued process refinement. Each process step must be monitored by means of human or automated visual inspection. During the initial start-up of a product, 100% inspection is not uncommon. After process stabilization, only sampling of the assembly is required. A sampling procedure, for example, would randomly select one or several units at a specific interval of time. As each unit is inspected, defects exceeding standard limits are identified and recorded.

Defect monitoring stations for stencil-applied solder paste and reflow-solder processing will include each of the following process steps; solder paste application, component placement, and reflow-solder processing.

153

Concerns and process refinement

The defect ratio from one assembly to the other does not seem to be affected by the reflow process as much as the solder paste characteristics. The viscosity of the solder paste will have a more significant impact on the solder joint defect ratio of the 0.4–0.5-mm (0.016–0.020-in) devices.

Although solder volume of each solder joint can be mathematically modeled, visual inspection on the finer pitch devices is not without compromise. A 100% inspection of every lead of the fine-pitch device in volume production is not practical and the use of advanced inspection systems for fine-pitch devices is inevitable. For example, ultrasonic imaging, X-ray, and X-ray laminography might prove to be very effective in preforming a nondestructive solder quality certification to measure solder density or detection of solder voids.

Solder paste stencil development

Several methods for depositing solder paste are available for surface mount assembly, but stencil application is most often the choice, especially for in-line assembly processing. Two areas of concern when auditing the solder stencil operation are alignment of the transferred paste material to the substrate surface and consistency of the paste volume on the surface mount land pattern. Defects can be controlled through refinement of the stencil fixture. Solder paste registration for the majority of the surface mount devices can be improved, for example, by reducing the overall stencil opening, and to compensate for all the tolerance variables of a typical PC board, a reduction of the stencil opening by 10–20% might be adequate. See details in Fig. 7-4.

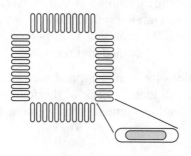

■ **7-4** *Solder paste registration for the majority of the surface mount devices can be improved, for example, by reducing the overall stencil opening, and to compensate for all the tolerance variables of a typical PC board.*

Fiducial targets for stencil alignment

To accommodate precise registration of the stencil for fine-pitch surface mount devices, some means of vision or camera-assisted alignment is required. For those systems using automated vision alignment of stencil to board, the board designer must furnish at least two global fiducial targets within the pad layer design file.

The global fiducials are only partially etched into the top surface of the stencil material and filled with black epoxy. The stencil machine camera system will automatically align the board to the stencil pattern for precise solder paste transfer. See Fig. 7-5.

Placement accuracy and land pattern geometry

Less-than-perfect machine placement of passive and 1.27-mm (0.050-in) pitch surface mount devices can be tolerated to a limited extent. The land pattern geometry supplied in chapter 3

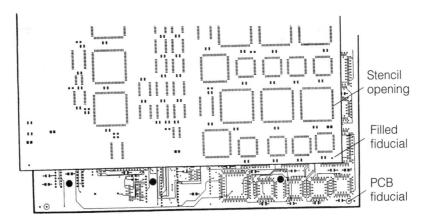

Stencil opening

Filled fiducial

PCB fiducial

■ **7-5** *The stencil machine camera system will automatically align the board to the stencil pattern for precise solder paste transfer.*

should accommodate both component tolerance and process variables. If you are unfamiliar with the reflow-solder process, note that during the reflow-solder processing, the entire assembly is heated to approximately 200° C and the solder paste is converted to a liquid state.

The devices are momentarily suspended in the liquid alloy and, through surface tension, the device mass will tend to self-center before the solder begins to cool. This phenomena, while predictable for passive and coarse-pitch devices, cannot be relied upon for fine-pitch attachment.

In the case of fine-pitch attachment, land pattern geometry is much more critical. The land pattern geometry must provide for both the minimum and maximum material conditions of the device and substrate. The designer must calculate the basic land pattern limits with provision for inspection and machine placement tolerances. The detail shown in Fig. 7-6 is typical of the tolerance factors for a fine-pitch device, placement accuracy, and the resulting land pattern.

Solder process evaluation

Because solder paste materials are not consistent from one supplier to another, the reflow-solder process profile can vary a great deal. Systems for solder paste reflow processing use various meth-

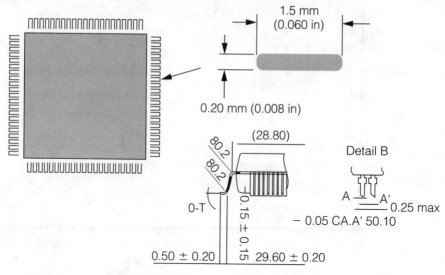

1.5 mm (0.060 in)

0.20 mm (0.008 in)

(28.80)

80.2

80.2

Detail B

0-T

0.15 ± 0.15

A ⊔⊔⊔ A'

0.25 max

− 0.05 CA.A' 50.10

0.50 ± 0.20   29.60 ± 0.20

■ **7-6** *The land pattern geometry developed for this fine pitch device allows for all material tolerances and will furnish the assembly process specialist the opportunity to meet J-STD-001 requirements.*

ods to generate heat or energy. In-line systems for instance, might use focused or nonfocused infrared energy, fluorinert vapor, convection hot air, or nitrogen. Some even use a combination of two technologies.

The designer of the PC board is not always cognizant of the final assembly process used, but a good design and correctly specified substrate will furnish manufacturing specialists with the opportunity to achieve satisfactory results from any one of these solder systems.

Inspection of the finished solder joint is generally a manual process using various degrees of magnification, but automated inspection methods can be adapted as well. Maintaining adequate clearance and consistent spacing between devices will ensure access for visual and automated solder inspection. A word of caution for users of the fine-pitch devices, the devices are susceptible to damage because of their relatively delicate leads. Avoid physical contact with the fine-pitch device after it is placed in the solder paste. Precise placement and alignment of the device to the land pattern is paramount to achieving a defect-free process.

Cause and effect of solder defects

Solder defects identified on fine-pitch devices following reflow processing and cleaning can often include the following:

☐ *Solder bridges* Excessive solder spanning from one contact to the other.

☐ *Insufficient solder* Cold solder connection or excessive void in solder joint.

☐ *Assembly error* Misalignment or missing parts.

Solder bridge or shorts between leads are often process related. However, the PC board design can also be a factor. Exposed signal traces near a land pattern are susceptible to bridging from solder balls or other particles. To avoid this occurrence, the designer must ensure that signal traces are covered by the solder-mask coating. Solder bridging often occurs between fine-pitch leads when the solder paste is applied too thick. The correct stencil opening and thickness will regulate the solder volume reducing the occurrence of defects.

Defining solder mask requirements

Most surface mount PC boards are coated with solder mask, but solder mask coating if greater than 0.04-mm (0.0015-in) thick can affect solder-paste application. Surface mount PC boards, and especially those using fine-pitch devices, require a low-profile, photo-imaged solder mask.

The mask material must be applied with either a liquid (wet) process or laminated as a dry film. Dry film solder mask material is supplied in 0.07–0.10 mm (0.003–0.004 in) thickness and might be suitable for some surface mount products, but this material is not recommended for fine-pitch applications. Few companies supply a dry film that is thin enough to meet fine-pitch criteria, but several suppliers can furnish liquid photo-imaged solder mask.

The solder mask's role in controlling the solder defects during the reflow process is significant, and designers of the PC board should be aware of the proper clearance or air gap used to separate the land pattern features from the signal trace.

Although most fabricators will advise otherwise, the solder mask opening can be line-to-line with the land pattern. A nominal clearance is recommended to ensure efficient imaging during fabrication. In general, solder-mask openings should be 0.15 mm (0.006 in) greater in size than the land pattern feature. This allows a nominal 0.07-mm (0.003-in) clearance on all sides of the pads. To ensure that the signal trace is not exposed, the minimum clearance between land pattern features and conductor should be no closer than 0.12 mm (0.005 in). Figure 7-7 illustrates the details.

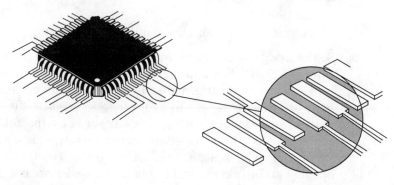

■ **7-7** *Solder mask between fine pitch contact land patterns is often preferred but as the lead pitch becomes smaller, maintaining solder mask features is more difficult.*

Although it is preferred that solder mask separates all land pattern features on the board, the lead spacing and pad size of fine-pitch devices will require special consideration. While the nonpartitioned solder mask opening or window on a four-sided quad pack is acceptable, it might be more difficult to control solder bridging between device leads.

A limit can be reached as to when solder paste and reflow-solder processes will not be practical. Devices having a lead pitch less than 0.4 mm (0.016 in) for example, might require an alternative post-assembly attachment method to maintain process yield and minimize solder defects.

Land pattern development for fine pitch

The designer should avoid attempting to rely on only a nominal tolerance factor for fine-pitch lead locations. The minimum and maximum dimensions are developed often from data supplied by several component manufacturers. Furthermore, it is likely that all those limits will be seen if more than one supplier is approved for the same device type.

A word of caution regarding the conversion of the metric dimension to the decimal inch equivalent: Even though the metric factor is converted to a three-place decimal, rounding off to one thousandth of an inch in one direction will create rapid tolerance accumulation. For example, tolerance accumulated at as few as twenty locations on one side of a fine-pitch device having only 0.65 mm or 0.024–0.025 in spacing between lead centers will force the last lead completely off the pad.

Converting metric dimensions

To ensure consistent land pattern alignment, the designer must develop the CAD data base from the component center line. In addition, if the inch dimensioning is mandatory then the compensation must be made to accumulate tolerance of metric conversion through alternate dimensioning of each contact pad location. That is, if 0.0245 is the resulting factor of the metric conversion and the CAD software is limited to three-place decimal dimensioning, then alternate the fourth decimal by one half thousandth of an inch from one land pattern to the other (0.024 in, 0.025 in, 0.024 in, 0.025 in), and so on. Examples comparing the impact of accumulated tolerance errors are illustrated in Fig. 7-8.

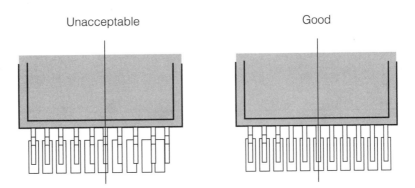

Unacceptable   Good

■ **7-8** *Accumulated tolerance of land pattern pitch will cause excessive misalignment of fine-pitch device leads.*

Providing for
test automation

WHEN A PRODUCT GOES INTO A HIGH-VOLUME PRODUCTION mode, automated testing of printed circuit board assemblies can prove to be the most efficient way to keep pace with the high-speed machines that produce them. The surface mount and fine-pitch assembly, unlike the pin-in-hole (PIH) assembly, must be planned for testability.

Assemblies with leaded devices or PIH technology provide a "test node" contact at every solder connection. Surface mount components are attached to land patterns and must rely on via holes or dedicated test pads for probe access. To regain 100% test access for the surface mount assembly, dedicated probe contact patterns must be included during the design layout phase of the product.

161

Planning for automatic testing

Preparing for automatic testing of assembled PC boards begins in the initial planning stage of design. The type of testing is determined by a number of factors, including volume, product life, and test equipment available.

Planning for testing includes agreement and understanding of several key factors including:

☐ Test methods.

☐ Test probe design options.

☐ Test probe contact size and spacing.

☐ Fixture tolerance allowances.

In the following text, surface mount assembly test options are explored, including process steps and the design guidelines that make the SMT assembly most efficient and economical to manufacture and test.

Many test levels are possible, depending on the environment in which the product operates. Three test methods popular for commercial and consumer applications are:

☐ *In-circuit test (ICT)* Making test probe connection to each common connection of two or more components on the board.

☐ *Functional cluster* Partitioning the PC board in a modular fashion with test points focused on the input and output of a circuit. Probe contact can be outside the component area.

☐ *Board level* Functional test using very refined test programs. With functional testing the board is analyzed through the connector interface using external or built-in diagnostics.

If electrical testing of the assembly is to be a part of the manufacturing process, the PC board design must accommodate the test strategy selected (in-circuit test, functional cluster, board level functional test).

In-circuit test

In-circuit test (ICT) is often used in the volume assembly operation as a means of verifying both assembly process control (identifying process defects) and exercising the function of each device in the circuit. Spring-loaded probes make electrical contact at each net or common connection within the circuit.

Using specialized software tools developed for the test system, every active lead of the IC (if accessible to probe contact) can be exercised through a custom test fixture and program. Test probing each contact point of every device is not a requirement when they are part of a common network; however, additional probes can be added to a network for measuring specific characteristics. Each test fixture is unique to the individual circuit board, and electronic software and hardware must be developed for each assembly. To make sure the board aligns perfectly to the probe contacts, the design of the fixture will also include tooling pins for positioning the assembly.

During ICT, if a device does not perform as specified, or an assembly process defect is detected during the test, a report is produced by the testing system that locates the process defect or directs the replacement of the failed part.

Requirements for ICT

The PC board designer and test engineer should define the test requirements before the design is started. For example, the test engineer might want provisions for adding pull-up or pull-down resistors on input and output lines. The connection of unused gates and control lines to power (VCC) or ground through a resistor is preferred and the designer should make provisions to probe all unused active pins of an IC. By accessing all contacts, each device function can be thoroughly exercised or measured during ICT.

ICT checklist

The following can be used as a checklist for the ICT test:

☐ One test node per net.

☐ Probe spacing of 2.0 mm C-C (0.080 in) minimum.[1]

☐ Contact area of 0.9 mm (0.035 in) minimum.

☐ All node access from one side.

☐ Probe/device clearance of 0.8 mm (0.030 in).

☐ Provide test node on (active) unused IC pins.

[1] Probe contact pins are available for 1.27-mm (0.050-in) spacing; however, these probes are not as reliable as the larger probes.

Preferably, probe contact is provided on the secondary side.

Function cluster test

Some circuits can be tested efficiently as a function. That is, only input and output signals of several interrelated devices are accessed to measure the performance of the circuit function. Typically the test probe contact at a common node of several components is more practical for locating component or assembly-process defects, but physical access for probing each net might be restricted.

The electrical test of a partitioned cluster or function can speed up the test cycle and isolate problem areas, but a specific defect location cannot be automatically identified. Instead, a technician will probe the suspected area manually until the problem has been identified. Because each network is not accessed, automatic measuring of device level functions or the value of individual components is not possible.

Functional cluster testing of an amplifier circuit, for example, works best by driving a defined low-level signal at the circuit input and measuring the amplified output level of the circuit. Several functional clusters can be measured simultaneously using a dedicated probe fixture typical of that developed for in-circuit tests, but probe locations will be limited to only input and output locations.

Board level functional test

Board level testing might require more sophisticated diagnostic programming to isolate the trouble area, but even the most efficient functional test might not detect a fault at the component level. Functional testing at board level typically uses test fixtures and software dedicated to the specific assembly under test. Measuring equipment and product-focused engineering is more specialized, but the test fixture hardware is generally far less complex than ICT fixtures. When manufacturing yield reaches a consistent process defect-free level on a product, ICT can be bypassed altogether. When this strategy is adapted, only those few assemblies that cannot pass functional testing are passed through the in-circuit test cycle.

Test probe design options

Spring-loaded test probes are supplied by several manufacturers. Several probe tip styles are available and are generally interchangeable in the holes of the test bed fixture. Options available for the probe tip will accommodate most contact requirements for surface mounted and PIH assemblies. The test engineer must select the length and tip design needed for the specific application. Examples of several standard probe contact (tip) designs are shown in Fig. 8-1.

A practical probe tip design for PC board surface connection can mate with either a flush contact area or a via hole pad designated as a test point. For example, the "chisel" point design just illustrated will provide for a positive electrical connection to a via pad, but the "crown" type might be best suited for flat-pad contact. Because the leads of PIH devices are often bent or crimped away from the center of the hole, the probe design must contact an area equal to the limits of the bends. Making contact to the lead of the PIH device will require a probe tip that can connect to somewhat inconsistent targets.

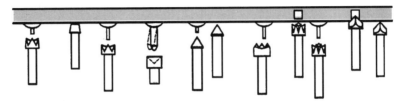

■ **8-1** *Test probe styles have been designed to make contact with most lead types or test pad configurations.*

Test contact size and spacing

Probing of the finished assembly using automated test systems is generally made on the secondary or lowest profile side of the board. The leaded part will always be accessible for probing because the leads will protrude through the board from the primary side. The SMT design will require accessibility to a via hole or other features reserved for test probe contact. Details are shown in Fig. 8-2.

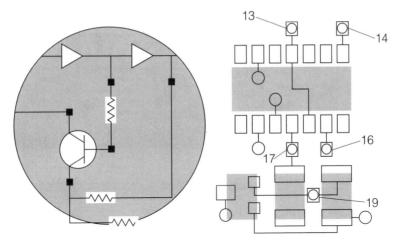

■ **8-2** *The designer must provide for test probe access to every "net" or common connection between device contacts. Test probe contact is generally made on the surface opposite the side with the greater component population.*

Clearance recommended between probe centers is 2.0 mm to 2.5 mm (0.080 in to 0.100 in) for standard probe contacts. Smaller probes allow for spacing as close as 1.3 mm (0.050 in), but the price of the smaller probes is high and these more delicate contacts might compromise test reliability.

Test point contact size and spacing

The test point or pad size must accommodate the combined tolerances of the PC board and test fixtures. When adding all tolerance limits of the PC board and test fixture elements, one can conclude that a contact area of 0.63 mm (0.025 in) diameter would be adequate. However, larger probe areas 0.9–1.0 mm (0.035 –0.040 in) diameter or square, for test points would be more reliable.

The test point contact should be located clear of the component body and land pattern area, because a test probe contact on the uneven or tapered solder joint cannot be tolerated. Any side loading of the fragile test probe might damage the spring action required for proper contact pressure.

The illustration in Fig. 8-3 represents a typical circuit with a test pad for each net or node. Notice that all test points are clear of the component body, thereby reducing the danger of mechanical interference.

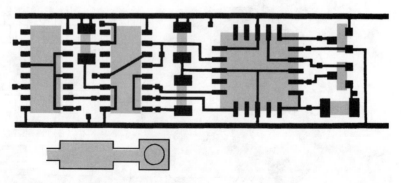

■ **8-3** *Test probe contact size and location must allow unobstructed access. Probe tips might damage the fragile ceramic materials common to resistors and monolithic capacitors.*

Test probe contact identification

The test engineer should be involved with the identification and numbering of test nodes on the schematic prior to the PC board layout. If the probe location and net is defined in the design, the X - Y position of each test node can be furnished for test fixture and program development. In addition, the test engineer will need a detailed report that furnishes net identification and a reference designator for each component. This data is vital in preparing test fixtures and the development of test software.

If devices are attached to the secondary side of the PC board assembly, the test point contact must be well clear of any component body. To easily distinguish the test pad area from other via pads, consider using a different size or shape pad. For example, a round via pad could be used for general front-to-back interconnections and a larger or a square pad could be reserved for test probe locations only, as shown in Fig. 8-4.

When developing the CAD data, a layer might be reserved to provide for test contact identification. Because a number is assigned to each net, the test point can be defined as a component within the net and verified as a specific physical feature.

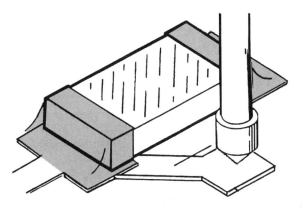

■ **8-4** *The designer must provide a probe contact area that will compensate for the tolerance accumulation of the test fixture parts and a typical fabricated substrate.*

Tooling holes for fixture alignment

Mating the completed assembly to the test fixture requires a minimum of two guide pins. The tooling pins act as a positioning guide to assist in the precise alignment of the probe contacts. The tooling holes provided for fixture alignment of the assembly should be at diagonal corners of the assembly. The hole size recommended for mating the assembled board to a fixture is 2.5 to 3.1 mm (0.100 to 0.125 in) diameter, free of plating. To avoid damaging devices in and around the tooling pins, a clearance of 3.1 to 6.2 mm (0.125 to 0.250 in) from the tooling pin body to adjacent components and leads is also recommended. See the details in Fig. 8-5.

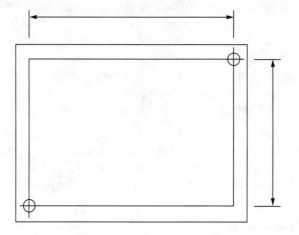

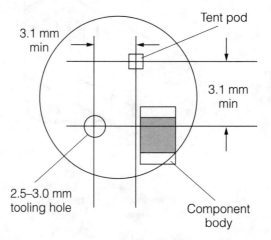

Tent pod

3.1 mm
min

3.1 mm
min

2.5–3.0 mm
tooling hole

Component
body

■ **8-5** *To avoid damaging devices in and around the tooling pins, a clearance of 3.1 to 6.2 mm (0.125 to 0.250 in) from the tooling pin body to adjacent components and leads is also recommended.*

Component density is often very high on surface mount assemblies and providing probe access during board design might take extra time and ingenuity. Even though SMT applications have higher component densities, the designer should make every effort to develop a board that will accommodate test probe access from one side.

When probe access is restricted on the secondary side, a two-sided test fixture can be developed to contact the primary and secondary sides of the assembly simultaneously. Two-sided fixturing is extremely complex and should be employed only as a last resort. See Fig. 8-6 for details.

Test fixture complexity also impacts reliability and development cost. The cost of two-sided fixtures can exceed three times the cost of a single-sided fixture. As an example, a two-sided test fixture developed for a commercial lap-top computer assembly would cost approximately $15,000.00, but if the design of the

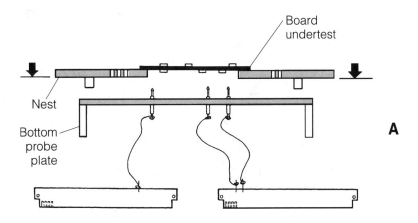

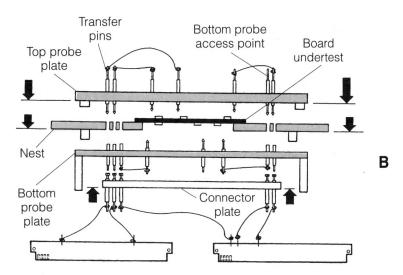

■ **8-6** *(A) Testing of SMT assembly. (B) Two-sided testing of SMT assembly.*

board allowed access to all nets on one side, the cost would be closer to approximately $5,000.00.

Spacing between probe contact pads should be as generous as possible. In addition, the designer should always avoid clustering too many probes in a small area. The force developed by the excessive density of spring-probe contacts can distort or damage the assembly. An even distribution of probe locations will reduce excessive force and ensure consistent probe contact pressure throughout the board. Probe contact locations should be well clear of component bodies as shown in Fig. 8-7.

- PC board tooling hole location   +/− 0.005 in/0.010 in mm
- PC board tooling hole diameter   +/− 0.001 in/0.003 in mm
- Component body dimensions   +/− 0.005 in/0.010 in mm
- Test probe location in fixture   +/− 0.005 in/0.010 in mm

■ **8-7** *To prevent damage to ceramic devices attached to the secondary side of the board, test probe contact features must be positioned away from the conponent body.*

Design for test checklist

☐ Divide complex logic circuits into smaller, functional logic sections.

☐ Use a single, large-edge connector to provide input/output pins and test/control points.

☐ Provide adequate decoupling at the board interface and locally at each integrated circuit.

☐ Never exceed the logic rated fan-out; keep it to a minimum.

☐ Terminate unused logic pins with a pull-up resistor to minimize noise pickup.

☐ Avoid using redundant logic to minimize undetectable faults.

170

- [ ] Construct trees to check the parity of selected groups of eight bits or fewer.
- [ ] Interrupt the circuit when a logic element fans out to several places that converge later.
- [ ] Standardize power and ground pins to avoid test-harness multiplicity.
- [ ] Provide a means of disabling on-board clocks so that the tester clock can be substituted.
- [ ] Group together related signal lines of particular families.

Test fixture preparation

Tolerance capabilities of precision machine technology and commercial PC board fabrication are quite different. It is common for PC board shops to profile the etched circuit with high-speed cutters using numerically controlled routing systems. Even with the use of the NC router, the accuracy of outside dimensions of the PC board in relation to the tooling hole center could vary by up to 0.25 mm (0.010 in).

Some companies might consider die punching the PC board using hard tooling for the high-volume or low-cost, paper-based epoxy boards. The cost of this tooling is far greater than programming for NC routing but the accuracy and price per board in high volume can easily compensate for the expenditure.

When developing a test fixture, hardware is often prepared using more accurate machining methods than used for board fabrication. Machine tooling and fixtures are a more accurate process and afford closer tolerance capability. Knowing what is possible and what to expect from test fixture services will help determine the location accuracy of the test probe contacts. Probe contact location accuracy is determined by the physical stability of base material used to retain the probe contacts and the equipment used to prepare the test fixture. The detail shown in Fig. 8-8 will illustrate the tolerance accumulation of the test fixture and a typical SMT circuit board.

To avoid excessive costs for testing and fixture development, the PC board designer and test engineering specialist must define the test strategy. Design for testing the surface mount product at the beginning of the program is vital, because adding testability to the board after the design has been completed can be difficult and compromises are often made to facilitate the limited access to via holes or test pads.

Specifying substrate materials and fabrication options

Base material selection for PC boards

GLASS-REINFORCED EPOXY RESIN LAMINATE (FR-4) HAS been the primary substrate material for etched copper circuit boards for many years. FR-4 has proven ideal for two-sided and multilayer PC boards that use plated through-holes to interface one layer with another. The fire-retardant epoxy glass material has become the industry standard for communications, instrumentation, computers, and virtually all electronic products requiring high quality and performance.

FR-2 low-cost, flame-retardant paper/epoxy composite is the workhorse of consumer electronics when electrical and mechanical properties are not overly demanding.

FR-4 fire-retardant glass/epoxy laminate is more expensive than the above materials, but widely used in computer and telecommunications applications.

FR-6 glass-mat, reinforced polyester is a low-cost favorite of the automotive industry, used for dashboard electronics and radios.

Although these materials are widely used throughout the industry there are alternatives to FR-4 in electronic applications. Low-cost consumer product manufacturers can use one of several paper-based laminates on the market. Etched circuits with simple functions might require copper conductor traces on one surface, with paper-based material holes for component leads drilled or punched through the substrate as shown in Fig. 9-1.

CEM-1 paper-glass composite is the best laminate property for price ratio. It can be used to replace other laminate grades to im-

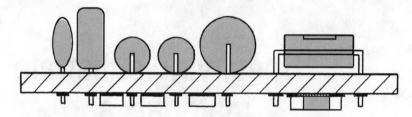

■ **9-1** *Low cost etched circuits having relatively simple functions may use paper based epoxy laminate and require only one circuit layer for interconnecting both leaded and surface mount devices.*

prove yields, cut laminate costs, or obtain better process. (For example, it can replace single-sided FR-4.) It is used extensively in radios, smoke detectors, and other consumer applications.

CRM-5 composite material uses a polyester resin system to bind together a sandwich of random glass mat core and woven glass fabric surfaces.

CEM-3 all glass composite is a recent entry into the laminate spectrum. It is cheaper than FR-4 with almost equivalent properties. It is entering automotive and appliance applications.

High-performance PC board materials

In addition to the commercial materials listed, the designer/engineer can handle more demanding environments with one of the "high-tech" materials available. Teflon and ceramic substrates are used in many RF applications requiring a stable dielectric characteristic. The Teflon materials are furnished with copper-clad for etched circuit requirements as are the more conventional laminates, and Teflon boards are processed in a method similar to FR-4.

Laminated material for SMT and MCM-C

Glass-reinforced laminate is the primary substrate material for etched copper, multilayer circuit boards for both MCM and SMT applications. Many alternative high-grade materials exist for use in electronic applications. Alternative materials to the FR-4 are widely used for high-performance surface mount products and can provide the designer with more options in both fabrication complexity and physical stability during assembly processing.

Most commercial SMT and MCM-L products will continue to use high-grade "FR-4 multifunctional" and "FR-4 tetra-functional" glass-reinforced epoxy laminate, while paper epoxy will be confined to consumer electronics. The following are a sampling of resin systems, reinforcement materials, and thermal-management materials available for printed circuit board fabrication:

Standard and specialized resin types	Tg temp.
☐ General-purpose FR-4 epoxy	130°C
☐ Tetrafunctional	137°C
☐ Multifunctional	150°C
☐ Bismaleimide Triazine (BT) epoxy	180°C
☐ Cyanate Esters	250°C
☐ Polyimide (non-MDA)	270°C

Reinforcement materials

☐ E-Glass   ☐ Aramids

☐ D-Glass   ☐ Kevlar

☐ S-Glass   ☐ Quartz

Thermal management alloys

☐ Copper (best thermal conductor).

☐ Aluminum (good thermal conductor).

☐ Copper Molybdemum Copper (physically stable, but not the best conductor).

☐ Copper Invar Copper (very stable, best TCE ratio).

High-performance laminates

Alternatives to the more common laminates, noted in Table 9-1, are polyimide glass, polyimide quartz, and Cyanate Ester materials. More stable than FR-4 materials, the polyimide products can be processed with most of the same techniques used in standard board fabrication. Because the polyimide and Cyanate Ester have a more stable thermal coefficient of expansion than FR-4, they are popular for chip-on-board applications typical of those used in MCM circuits. Both polyimide and Cyanate Ester can be laminated to metal core, like copper-invar-copper layers to provide an even more thermally stable substrate surface, but the height and weight of the reinforcement material is a significant factor to consider.

■ Table 9-1 Dielectric constants and thermal coefficients of expansion for polyimide and related materials.

Material type	Dielectric constant	Thermal coefficient of expansion
Epoxy glass	5.0–6.0	15.8
Polyimide	4.3–5.0	14.2
Polyimide/CEC*	4.3–5.0	6.4
Ceramic (alumina)	8.5–9.5	6.4
Epoxy Kevlar	3.8–4.5	6.5
Epoxy Quartz	3.5–4.0	6.0

*Copper invar copper metal core.

When a designer must meet special requirements for board material, contact the board fabricator for assistance in developing the specifications. This will help to eliminate any issues that could delay the processing of the order.

Ceramic Multichip Module MCM-C

Ceramic substrates are fabricated using very different techniques than laminated circuits. The substrate size is limited and conventional drilling or routing of this material is not possible. The multilayer ceramic substrate is generally punched and profiled in its green or unfired state. After firing of the organic material, holes and additional alterations to its shape require laser technology.

The circuit traces are added to the surfaces of the ceramic layers, and conductive fillers can be drawn into the laser-drilled via holes to connect one layer to the other. Ceramic substrates are used primarily for hybrid circuits. In most cases, unpackaged IC chips are attached directly to the surface of the ceramic and terminated using wire bonding (COB) techniques.

The advantage in using a ceramic for the circuit is the choice the engineer has in mixing several integrated circuits, transistors, and miniature devices into a very small area. The result is a customized electronic function in a small package. The finished multichip module (MCM-C) can have leads attached directly to contact pads around the substrate edges for mounting into a larger PC board or, as in hi-rel applications, hermetically sealed in a metal housing with glass-insulated leads. See Fig. 9-2.

Contact the ceramic substrate fabrication specialist to assist in writing specifications. A clear understanding of the requirements

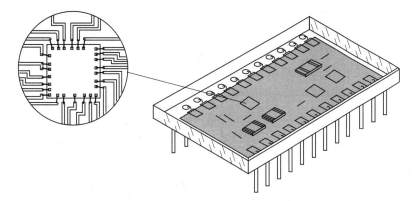

■ **9-2** *Hybrid conductive circuits are added to the ceramic substrate and components terminated with solder, conductive epoxy, and wire-bond technology.*

will eliminate misunderstandings, which could delay the processing of the product.

Fabrication and material planning

The majority of commercial electronics will continue to use one of the epoxy-glass laminate systems, while paper-epoxy will be confined to consumer products. Substrate materials are generally furnished to the board fabricator in large 36 × 48-inch sheets, although size will vary in different parts of the world. The fabricator usually cuts the sheet material into smaller panels for processing, i.e., 18 × 24 inches, or 12 × 18 inches. From these panels, the PC board is processed into one, two, or more PC units per panel. The fabricator, in an attempt to keep costs under strict control, should plan material usage carefully for minimum waste.

When designing the board, furnish at least two unplated holes for tooling location pins on each unit. If boards are furnished in a multiple unit panel, two additional tooling holes are required, generally on a break-away tab or outside the individual circuit board area. If the designer of the board prepares film in multiple image format, two areas are to be considered: the maximum size board the assembly equipment can handle, and minimizing waste of the board material. See Fig. 9-3.

The material outside the board area will be discarded after assembly processes are completed. Tooling holes furnished within each board will be used for holding the boards during numerically controlled (NC) routing or later in test fixtures for the finished as-

177

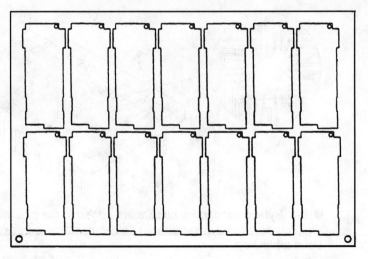

■ **9-3** *Small and irregular shaped PC boards may be arranged in a panel array format but the finished size of the panel must consider the fabricators working panel size to minimize material waste.*

sembly. When additional holes are furnished on the edge of a panel, each can be round or specify one round and one oblong. The optional elongated hole allows the assembly to be extracted from the tooling pins without binding. The tooling hole clearance must allow for placement of SMDs on both sides of the board.

While machine NC routing to profile the board is one of the last steps in the fabrication process, planning must be done in advance to ensure the best results. Methods of profiling the individual board shape include rotational cutters, high-pressure hydro routing, laser profile, bevel scoring and shearing after assembly. For the latter, adequate clearance to component bodies and circuit traces is necessary for straight cuts.

The slot separating the breakaway tab from the individual boards on a panel should be a minimum of 2.5-mm (0.100-in) wide. Generally a 2.3-mm (0.090-in) diameter router tool is used to create the full width of the slot in two passes. One pass of the router tool will be programmed to precisely cut the edge of the board unit outline and the second pass of the cutter will profile the other board unit. Precise control of the board edge routing will help to ensure greater tooling hole-to-board edge accuracy. The routed slot pattern between each board unit is interrupted by strategically located sections of solid PC board material or "tabs". The connecting tabs, typical of those shown in Fig. 9-4, allow several substrate images to be combined onto one panel for efficient automated assembly machine handling.

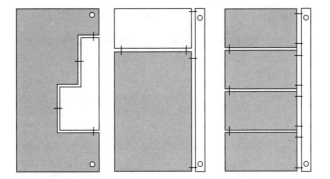

■ **9-4** *The connecting tabs allow several substrate images to be combined onto one panel for efficient automated assembly machine handling.*

Fabrication options

Tab sections retain each board in the panel array. Tabs retaining the board to the panel are usually 2.5 mm (0.100 in) or wider and are spaced to adequately support the panel as a unit through the assembly process. There are several variations used for the connecting tabs. Figure 9-5 illustrates a few.

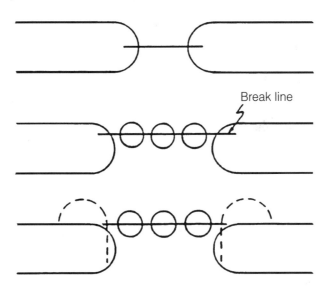

■ **9-5** *Connecting tabs between smaller boards or edge extensions can be a solid web section or drilled in a pattern to relieve stress at the break point.*

Panel planning

Several techniques for retaining the substrate in a panel form are available. Each of these examples represents an optional routed slot pattern.

Die cutting or punching methods are normally reserved for higher volume products and have proven to be efficient means of fabrication for paper-based laminates, while glass laminates are less than ideal for this method, due to the abrasive nature of the glass fiber material. Punch-press profiling of the board (Fig. 9-6) can be configured similar to a routed panel, but processed with greater speed and economy. Panelizing small boards is a must for efficient assembly processing.

Figure 9-7 illustrates an example of an irregularly shaped board that is awkward to handle as an individual unit.

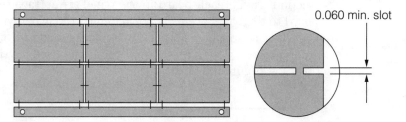

0.060 min. slot

■ **9-6** *Die punched slot and tab methods require a hard tooling investment, but can reduce panel fabrication costs significantly over the life of a product.*

Dimensioning the PC board substrate

Locating and dimensioning of tooling holes and other physical features of the substrate should follow established guidelines. As a resource to designers preparing detail fabrication drawing for the printed circuit board, refer to ANSI/IPC-D-300G—*Printed board dimensions and tolerances*. Included in the standard are recommendations and established industry limits on material and processes.

Multilayer and fine line

The industry is pressed to push the state of the art in fabrication technology as component density increases and circuit complexity evolves. When the trace width and air gap are less than 0.2 mm (0.008 in), or plated through-hole diameters are less than 0.5-mm

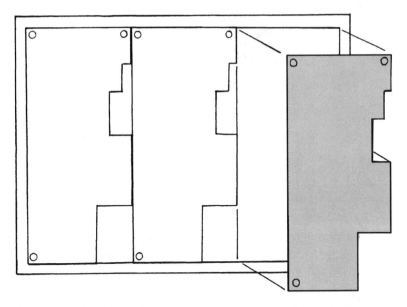

■ **9-7** *Panelized boards can be cut from the panel format with a precision router following all assembly processing.*

(0.019-in) diameter, the increased difficulty of manufacturing will be reflected in the board cost. Before reverting to the more difficult and costly fine-line method of board manufacturing, the designer should make an effort to keep fabrication less complex.

For circuit boards requiring the greater circuit density, the designer is advised to confine the fine-line circuit traces 0.13–0.15 mm (0.005–0.006 in) to internal layers. The board is usually one of the most expensive components of the assembly; therefore, savings at this level will be reflected over the entire life of the product.

Via hole planning

To maximize drill speed and keep the drill breakage rate low, most board specialists would prefer a minimum finished hole size for 1.5-mm (0.062-in) thick boards of 0.5-mm (0.019-in) diameter for the SMT substrate. This 3:1 size ratio allows for a reliable plated hole and increased conductor trace density. Standardizing on the 0.5-mm finished hole diameter will allow for a 1.0 mm (0.040 in) outside diameter pad, maintaining the desired 0.25 mm (0.010 in) annular ring. See Figs. 9-8 through 9-10 for details.

Some SMT applications might have even higher device and conductor density requirements. Smaller via holes and pads can be furnished to meet these requirements, but both cost and produc-

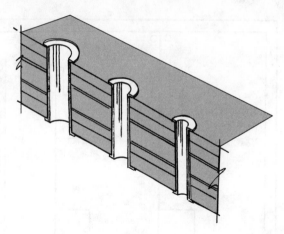

■ **9-8** *When planning circuit density, consider the impact of the aspect ratio between the plated via hole diameter and the finished board thickness.*

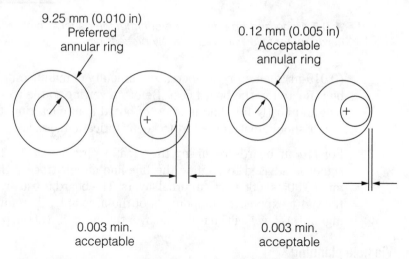

9.25 mm (0.010 in)
Preferred
annular ring

0.12 mm (0.005 in)
Acceptable
annular ring

0.003 min.
acceptable

0.003 min.
acceptable

■ **9-9** *Pad diameter for via holes must consider the fabrication tolerance limits and minimum annular ring requirements.*

tion yield will be impacted. A finished plated hole size of 0.3 mm is within the technology capability of many volume fabricators and for some thin materials, 0.15-mm diameter holes can be provided. Aspect ratio of plated holes in a printed circuit board substrate can impact cost as well as process yield and effect reliability.

Smaller via holes for high-density multilayer SMT and MCM-L

As circuits become more complex and component density is increased, the need to add plated through (via) pads in the sub-

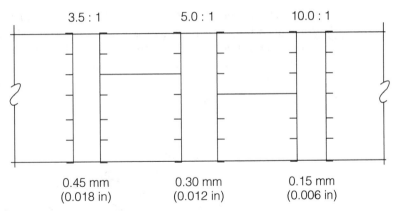

3.5 : 1	5.0 : 1	10.0 : 1
0.45 mm (0.018 in)	0.30 mm (0.012 in)	0.15 mm (0.006 in)

■ **9-10** *This 3:1 size ratio allows for a reliable plated hole and increased conductor trace density.*

strate to maximize available space also increases. Because plated through-holes will not have to clear a component lead, a much smaller drill size is possible. In glass laminate fabrication, very small drill sizes can add excessive cost to the fabrication of the PC board. High-density SMT and MCM-L designs generally require smaller via holes, and in some cases, blind and/or buried vias; however, the designer must expect the costs per square inch to rise dramatically.

Preferred plated hole diameter to finished board thickness: Finished board thickness to recommended hole size.

- ☐ 0.50 mm (0.019 in) thick or less, 0.30 mm (0.012 in) hole diameter.
- ☐ 0.56 mm (0.062 in) thick or less, 0.50 mm (0.020 in) hole diameter.
- ☐ 1.90 mm (0.075 in) thick or less, >0.63 mm (0.025 in) hole diameter.
- ☐ 2.50 mm (0.100 in) thick or less, >0.89 mm (0.035 in) hole diameter.

Providing for high-density circuit routing

To further provide for routing conductor traces and at the same time ensure an acceptable air gap, you might choose to use a square pad for via holes. The square configuration will furnish more than enough metal in the diagonal corners of the pad to compensate for the reduced annular cross section at the sides of the square. The square via pad can be closely spaced when necessary, or on an established grid, it is possible to route two or three conductor traces between pads as shown in Fig. 9-11. The reduced size of the plated via hole diameter and pad often will allow more conductor traces on internal layers. Small via pads and holes required for higher density SMT will accommodate higher circuit routing density.

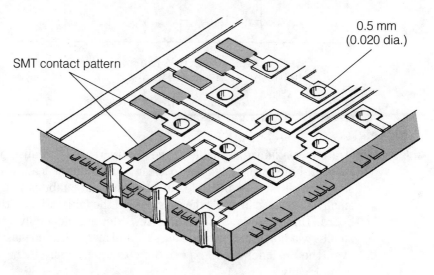

0.5 mm
(0.020 dia.)

SMT contact pattern

■ **9-11** *Component land pattern breakout to the grid spaced via hole and pad improves conductor trace routing paths and provides for solder mask separation.*

Plated through-holes for lead device attachment

Although the designer is advised to make every effort to use surface mount devices where possible, leaded through-hole parts continue to be a viable part of the electronics industry. Because leaded PIH devices have been used for more years than surface mount, the guidelines for sizing plated holes are well established.

Due to the difference in components, the lead diameters can vary over a wide range. To reduce drill size variables and fabrication cost, PC board manufacturers advise combining several size leads to one hole size when lead diameters are within 0.127 mm (0.005 in) of each other. For the majority of through-hole discrete components, the designer will find that lead diameters fall in the range of 0.45–0.7 mm (0.018–0.028 in). A common hole size usually can be selected if basic allowances are made for the tolerance of lead-forming equipment, automatic insertion accuracy, and wave solder characteristics. As an example, the industry has generally accepted the 0.94–1.07 mm (0.037–0.042 in) diameter hole range as a standard for auto-insertion application. Smaller holes can usually be adapted for low-volume applications using hand assembly.

The recommended hole provides 0.25–0.30 mm (0.010–0.012 in) clearance for the larger diameter leads supplied on resistors and capacitors, while at the same time compensating for the more difficult alignment when automatically inserting DIP ICs.

Manufacturers of fixed-space leaded components have complied with the 0.100-in grid arrangement. The 0.100-in grid pattern has been used for many years as a standardized form with PC board designers as well. It is a common practice to use a conductor trace between 0.15–0.20 mm (0.006–0.008 in) for most signal-carrying applications. Routing one or two conductor traces between 0.100-in spaced holes and packages will restrict the annular or outside diameter of the contact pad. The conductor path must be spaced away from nonrelated contact pads and conductors, maintaining an equal air gap or clearance between traces and other features. This air gap will allow for a clean etch during fabrication of the PC board and reduce the chance of solder bridging during assembly processes. See Fig. 9-12.

Outside layer   Inside layer

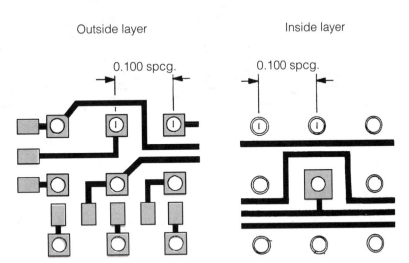

■ **9-12** *Conductor traces of less than 0.010 in. in width on internal layers can take advantage of the space occupied by the pad around each via hole on the outside layers.*

The tolerance of a drilled and plated hole is typically within 0.07–0.10-mm (0.003–0.004-in) diameter. Keeping the above-mentioned tolerances and limitations in mind, 1.27–1.60 mm (0.050 –0.062 in) outside diameter contact pad size is recommended for leaded components requiring the 0.100-in grid pattern.

Component density can be increased further when using multilayer technology because trace width and air gap on the inner layers will allow higher circuit routing density. Typically the via pad is retained on the outside layer, and on the inside layer only when a connection must be made.

CAD auto-routing of conductor traces is efficient when via holes are located on a fixed grid pattern. For example, two 0.15/ 0.20-mm (0.006/0.008-in) wide conductor traces will provide an equal air gap on these inside or laminated layers. The importance of accuracy in pad location and conductor trace spacing cannot be overstressed. Companies rely heavily on efficiency of routing software for CAD systems designed to alleviate these complex applications. Figure 9-13 provides an example.

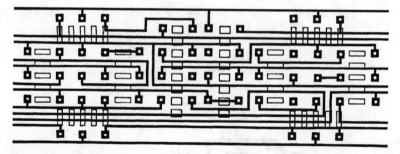

■ **9-13** *Transfering the circuit paths through via holes to the inner layers of the PC board will improve circuit routing effiency as well as increase component density.*

Higher circuit density for SMT

The surface mount component does not require a hole for each contact in the substrate. On a smaller, less complex surface mount circuit, it is possible to design the PC board without using any holes. As SMT circuits become more complex and component density increases, the need to reduce via pads and holes in the substrate becomes a necessity. Because via holes do not have to clear a component lead, a smaller drill size is possible.

In a glass laminate fabrication, an excessively small drill size will add significant cost to the fabrication of the PC board. To maximize drill speed and to keep the drill breakage rate low, most board shops would prefer a minimum finished hole size of 0.45–0.5-mm (0.018–0.020-in) diameter. This size of a hole allows for a reduction in pad size and increased conductor trace density. Standardizing on a 0.15-mm (0.020-in) finished hole diameter will allow for a 1.0-mm (0.040-in) outside diameter pad while maintaining the desired 0.25-mm (0.010-in) annular ring. See Fig. 9-14.

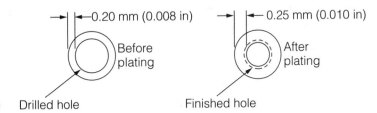

■ **9-14** *Standardizing on a 0.15-mm (0.020-in) finished hole diameter will allow for a 1.0-mm (0.040-in) outside diameter pad while maintaining the desired 0.25-mm (0.010-in) annular ring.*

Multilayer design considerations

Using multilayer for surface mount applications will dramatically increase density possibilities. Conductor traces of 0.15 mm (0.006 in) or less in width on internal layers can take advantage of the open space not occupied by the device land patterns filling the area on the outside layers.

The reduced size of the typical plated via hole diameter will easily allow three conductor traces on internal layers without resorting to fine line <0.12 mm (<0.005 in) traces.

As component density increases further, it might be necessary to limit the outer layer surfaces to the component's contact patterns, via pads, and a short conductor trace for connection. This technique is referred to as *pads only* and has all circuit traces buried on internal layers of the PC board allowing maximum component density.

Providing for CAD auto-routing of circuits

Auto-routing the circuit traces for SMT is most efficient when a via pad and a hole are retained on all layers of the multilayer board. To simplify the circuit routing procedure, a standard contact or land pattern for each type of device would include via holes positioned in a wider pattern away from the device mount site. For circuit routing efficiency, each active lead of the device is assigned a via pad. The device land pattern is connected to the via pad with a relatively short circuit trace furnishing a uniform grid spaced pattern. Only the land pattern and via pad with its connecting signal trace is retained on the outer surface, and all the signal traces as well as power and ground planes are buried on inside layers of the board. When all signal traces are contained on inside layers, more precise photo-defined etching processes are used in fabricating the layers. Photo-defined etched circuits provide an opportunity to produce a finer circuit trace width and a narrower separation or air gap between each circuit.

To ensure that solder materials remain on the land pattern areas during the reflow assembly process, solder mask (polymer) coatings are applied to both surfaces of the finished PC board. Although it might be more economical to eliminate additional fabrication process steps, photo-imaged solder mask over bare copper circuit traces (SMOBC) is recommended for surface mount applications. With a photo-imaged solder mask process, very precise clearance can be maintained. The clearance separating the solder mask material from land pattern, bonding sites, and other features is 0.13 mm (0.005 in), but finer spacing is possible with the lower profile materials.

Mask material must not overlap onto the surface mount land patterns. On the other hand if the solder mask opening is too wide around the SMT device land patterns, solder paste will migrate away from the land pattern during the reflow-solder process typically used in assembly. Figure 9-15 compares examples of solder mask clearance for surface mount PC boards.

The solder mask acts to contain the paste and to ensure that each component contact receives an equal amount of solder. Contact areas are usually connected to a via pad with a narrow trace. This isolation is necessary to allow for a dam of solder mask to stop the solder from flowing to and down the via hole. Solder mask coating over the via hole is acceptable if the via pad is too close to the contact area to provide for an adequate solder mask barrier. Designers might choose to cover or "tent" all via pads with solder mask.

Selecting solder mask material

Three types of photo-imaged solder mask are available. They are:

☐ *Liquid photo-imaged* The liquid material is applied by screen flooding or curtain coat. This material is very thin (1.5 mil) and low in cost.

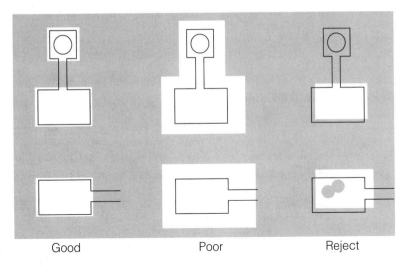

| Good | Poor | Reject |

■ **9-15** *Examples of solder mask clearance for surface mount PC boards.*

☐ *Dry film photo-imaged*   Applied with a lamination process, the thickness is 3 to 4 mils, which is too thick for fine-pitch applications.

☐ *Low profile, liquid/dry photo-imaged*   This material has two parts, a liquid coating followed by a very thin 1.5 mils film.

The low-profile (1.5 mils) mask requires specialized systems for application. If tenting or covering the via pad and holes is required, one or both sides can be covered. Don't cover pads or vias needed for test probe contact required for in-circuit testing of the finished assembly.

Plating options for SMT

One of the more process-compatible plating options for the surface mount attachment area is tin-lead. Tin-lead is a traditional plating for etched copper circuits, although tin-nickel is frequently used as well. The advantage of the tin-nickel surface is the flatness and uniformity of the finished surface. Other fabricators, to achieve that flatness, will strip the tin-lead from the board after etching, leaving bare copper exposed over all traces. As mentioned, solder mask is applied over the bare copper (SMOBC). Contacts and other features clear of the solder mask material are coated with tin-lead. The plating is achieved using a dipping process followed by hot-air leveling or selectively electroplated and fused. Dipping the board into a tin-lead solder bath followed by hot-air leveling adds the proper tin/lead alloy ratio to the contact areas for process compatibility. The same technique can be used on the nickel or tin/nickel plated board.

Specify standard materials

Specifying standard materials will ensure greater producibility, because this allows the fabricator to use "off the shelf" materials. Overall thickness can be specified including plating, but do not specify spacing between the layers created by the prepreg or core thickness.

If you must specify core and/or prepreg thickness or when dielectric spacing is critical to the performance of the circuit, allow the printed circuit board supplier a greater tolerance on the overall thickness to maximize fabrication efficiency and yield.

Preparing the printed circuit board fabrication detail

☐ Specify substrate materials.

☐ Identify layer count and sequence.

☐ Furnish copper thickness of all circuit layers.

☐ Specify dimensions for dielectric.

☐ Use datum dimensioning (examples shown in Fig. 9-16).

Other issues that must be clearly defined on the board specification notes are related to fabrication allowances and coating materials.

Furnish specific requirements

☐ Minimum lines and spaces.

☐ Minimum annular ring.

☐ Solder mask type and thickness limit.

☐ Plating or coating material.

☐ Legend or graphics color and type recommendations for developing the SMT substrate.

The following suggestions have been offered by board fabricators to assist the designer in planning the most reliable, physically stable, and cost-effective substrate possible.

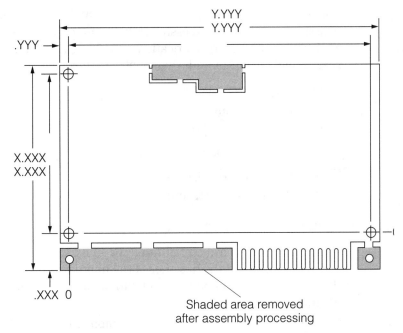

9-16 *To assure the mechanical accuracy of the finished PC board, profile dimensions should be referenced from a 0-0 datum hole or feature within the boards perimeter not from the boards edge.*

Symmetrical construction

☐ A conductor width of 0.2 mm (0.008 in) on outside layers.[1]

☐ A conductor width of 0.13/0.15 mm (0.005 in/0.006 in) on inner layers.

☐ Board thickness tolerance is 10% of nominal.

☐ Preferred 0.05 mm (0.002 in) minimum for dielectric thickness.

Maximize air gap between conductors and other features.

☐ Preferred board thickness to plated through-hole size is 3:1.

☐ Annular plating around hole is 0.13–0.18 mm (0.005–0.007 in).

☐ Minimum annular ring on plated holes is 0.13 mm (0.005 in).

☐ Conductor to edge: 2.54 mm (0.100 in) preferred., 0.80 mm (0.031 in) minimum.

---

[1] To provide equal stresses on both sides of the substrate at all times. Although preferred, 0.15 mm (0.006 in) is minimum conductor width and requires special processing.

General guidelines are often supplied by the PC board fabricator to assist in controlling excessive waste of material or costs associated with unnecessary fabrication complexity. The following issues should be considered when finalizing the design of the board:

Recommendations for economics

- [ ] Use standard materials
- [ ] Minimize hole size variation
- [ ] Plan material usage to avoid waste
- [ ] Allow for fabrication tolerances
- [ ] Provide a balanced structure

Planning efficient material usage

Fabricators will plan material usage carefully to minimize waste and control costs. If the designer of the multilayer board prepares the Gerber or IPC-D-350 file in a multiple image format, three areas must be known:

- [ ] The maximum size the assembly equipment can handle.
- [ ] The maximum size the laminating system can handle.
- [ ] Minimize excessive waste of the board material.

The United States-based fabricators cut the sheet material into smaller panels for processing. A typical panel size might be 14 × 24 in, 18 × 24 in, or 18 × 26 in (18 × 24 in. is most common). The individual PC board then would be processed as one, two, or more PC units per panel. When the board is designed for SMT, furnish a minimum of two nonplated holes for tooling pins on each assembly unit or panel.

Some SMT boards are too small to provide tooling holes, and when the board is developed as an array of smaller assemblies, tooling holes should be furnished on the panel perimeter. In the case of panelization, two additional tooling holes might be required for fixturing.

Panel planning for thin materials

Smaller SMT and MCM-L assemblies often require a thin finished board. For example, it is not uncommon for the PCMCIA substrate to be 6 to 8 circuit layers but less than 0.5-mm (0.020-in) thick. A panel array for the thin assembly must be engineered to allow for several assembly process steps without physical distortion. A panel configuration for the PCMCIA type product should be limited to no more than four units. A two unit panel might be preferred on complex assemblies having fine pitch on both sides, as shown in Fig. 9-17.

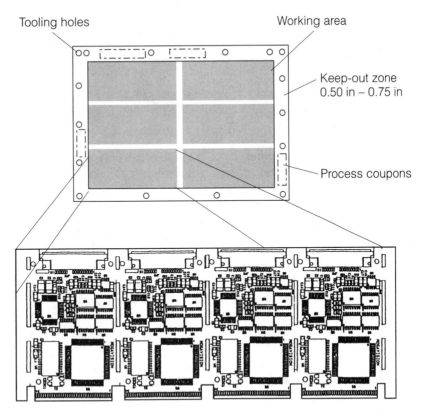

Tooling holes

Working area

Keep-out zone
0.50 in – 0.75 in

Process coupons

■ **9-17** *Multiple board unit layout must consider how the array will nest within a standard panel format for fabrication efficiency.*

The overall bare die attachment of IC size is limited for chip-on board. The most common wire bond systems are limited to a zone of 100-mm (4-in) square. Systems are available to process larger boards, but the decision to expand the panel to a larger size is not advised due to the cost impact of nonstandard wire bonding systems.

Plating requirements for COB processing

MCM-L assemblies designed for die attach and wire bond termination (COB) will require "bondable gold" plating over nickel at the wire bond sites. Gold thickness will vary dependent on the wire bond methods used for assembly.

Recommended bondable gold thickness:

☐ 35–50 micro inches for gold wire bonding
☐ 5 micro inches for aluminum wire bonding

193

Specifying substrate materials and fabrication options

The bonding wire is normally 0.025-mm (0.001-in) diameter gold or aluminum. The gold wire actually attaches to the gold surface plating so the thickness is much greater than that required for aluminum wire. The aluminum wire penetrates beyond the gold and onto the nickel plating.

Plating process for SMT

The process-compatible plating required on the contact area is tin-lead. Tin-lead is customary plating for etched copper circuits, although tin-nickel is frequently used. The advantage of the tin-nickel surface is the flatness and uniformity of the finished surface. To achieve that flatness, other fabricators will strip the tin-lead from the board after etching, leaving bare copper exposed over all traces. Solder mask is applied over the bare copper (SMOBC) and contacts are plated as a post-operation.

To be compatible with the surface mount devices and the solder and flux used, it is necessary to plate the contact areas and any plated through-hole and pad that mounts a leaded device with tin/lead. With the bare copper board, solder mask is applied, exposing only the contacts and holes. Dipping this board into a tin/lead solder bath with hot-air leveling adds the proper alloy to the contact areas for process compatibility.

The same technique can be used on the nickel-plated board. If solder mask is to be eliminated, electroplating of tin or tin-lead with a selective process must be used on contact areas and through-holes for mounting leaded components. In all cases, the tin-lead should be reflow-fused.

Typical fabrication notes for a two-sided surface mount PC board

Notes (unless otherwise specified):
1. Material 1: Type FR-4, Multifunctional ½ oz. to 1 oz. Copper clad. 0.062-in + 0.007-in thick.
2. Location of all holes to be within + 0.003 in. of pad center. All holes diameters are after plating.
3. Copper plating in all holes to be 0.25 mm (0.001 in) minimum.
4. Plating: (SMOBC, see note 5), unmasked areas to be 0.025 to 0.05 mm (1 to 2 mils) of 60/40 tin-lead solder using: plating, hot-air level. Solder covered areas to be flat and free of excess solder.

5. Coat both sides of the printed circuit board with photo-imaged solder mask. Align to true position within 0.002 in. Application of solder mask should be in accordance with IPC-SN-840, Type B, Class III. (All pads and contact areas to be free of solder mask.)

6. Space between trace and trace or trace and pad shall be within + 0.002 in of paths shown on film or CAD data.

7. Width of conductor paths and land pattern size shall be within 0.50 mm (0.002 in) of laser plotted photo-tool.

8. After baking in 150°C + 5°C oven for 15 minutes, board acceptability shall be based on IPC-A-600C, Class II.

9. Board twist and warp not to exceed 0.10 mm per 25.4 mm in length (0.004 in. per linear inch).

10. No voids, opens, or shorts allowed in traces.

11. Vendor code and UL rating shall be shown on circuit side of board.

12. Silkscreen legend to be applied to component side of printed circuit board using white epoxy ink. All pads must be free of legend.

13. Gold flash or plate finger tabs 1.2–2.0 micron, 0.00005–0.00008 in. thick over hard nickel.

Bare board test

Bare board test fixtures used to verify the quality of the pin-through-hole type circuit board probed only the plated holes connecting sides or layers of the finished substrate. This system detects breaks or openings in trace paths, bad hole-plating, and shorts.

In testing a board with surface mount components, the continuity of the plated holes is only part of the quality issue. To verify the circuit, the test probe contacts must be made at the end of every feature. A feature is the land pattern area that the SMT component lead-contact is attached to during the solder process.

These features, unlike the PIH boards, often have extremely close lead-to-lead spacing. It is common to have center-to-center spacing between leads of 0.050 in. on the small SOIC and PLCC ICs. The probes for testing these boards are individual spring-loaded contacts held in a custom-drilled polycarbonate panel fixture that aligns to the lead patterns of the devices, staggering pins on the rectangular footprint areas when required. With the addition of fine-pitch devices, 0.0315 in (0.8 mm), 0.025 in, 0.020 in, and smaller lead spacing, test fixtures have taken on a complexity new to the test process.

Using test block concepts of fine-wire probe clusters, a contact point can be made at each fixture. Test equipment manufacturers and even contract test houses are developing innovative techniques to provide the customer with reliable test data.

Base materials for flex circuits

The base material chosen to manufacture the flexible circuit must withstand a wide variety of process steps. The polyester material can tolerate process and environmental temperatures up to 200°C. This relatively low-temperature material is acceptable for many interconnection requirements or low-stress applications. The thicknesses of lower cost polyester can range from 0.002 to 0.010 in. Polyimide base materials, although more expensive, are tougher than the polyester and will withstand temperatures beyond 370°C. Polyimide base material thickness can be as thin as 0.001 in. or up to 0.005-in thick, with excellent dimensional stability. Base material of 0.002-in thickness or greater is preferred, however, due to the ease of processing and handling.

Copper foil lamination

Copper foil is press or vacuum laminated with 0.001-in acrylic adhesive to the base material. The foil is cut into manageable panel sizes of 12 × 12 inches or 12 × 24 inches, depending on the fabricator. Copper thickness ranges from ½ ounce to two ounces. The one ounce copper (0.0014-in thick) is used most often. Copper can be laminated to both surfaces of the base material, providing for two-sided circuits, or when necessary, laminated into multilayer substrates. These multilayer circuits, although less flexible, are processed in a similar fashion to the conventional rigid printed circuit board.

Single conductor layer flex circuit

Flexible circuits are developed through a progression of fabrication process steps. The materials include the base material layer, laminating (adhesive) film, copper foil, and cover material. Figure 9-18 illustrates the general fabrication steps in preparing a single copper layer flexible circuit.

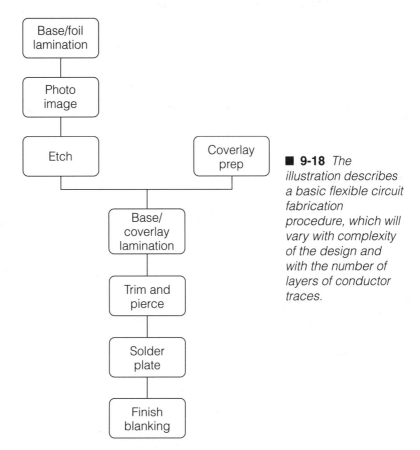

9-18 *The illustration describes a basic flexible circuit fabrication procedure, which will vary with complexity of the design and with the number of layers of conductor traces.*

A dry resist film is first laminated to the copper clad base material. Photo-tool artwork of multiple images of the circuit is often prepared on glass plates. The composite is "sandwiched" between glass circuit image masters. The glass master is a far more precise medium than a typical film photo tool used in conventional PC board fabrication. After exposure of the light-sensitive resist film on the copper, the exposed material is washed away, leaving a resist pattern over the proposed copper circuit. After curing the retained film image, the exposed copper area is chemically etched away.

Methods of etching differ from one fabricator to another, but generally, high-pressure spraying of the liquid etchant is used to remove the copper material. The copper traces and space between traces can be as little as 0.004 in. on ½ ounce and one ounce copper-clad material.

Specifying substrate materials and fabrication options

Coverlay preprocessing

Coverlay material is the same type of durable polyimide used for the base layer. The coverlay can be as thin as 0.025 or 0.050 mm (0.001 in or 0.002 in) and is applied to the base composite after etching the circuit pattern. All openings in the coverlay are predrilled, milled, or punched before lamination to the base circuit panel.

Coverlay lamination and piercing

The coverlay is "pin registered" to the finished base layer circuit and subjected again to the high-pressure lamination process. On production processing of the circuit, a "hard tool" die is used to punch-pierce holes or shapes into each of the circuit images on the laminated panels. Prototype or short-run panels will use milling or drilling equipment to provide openings or holes. This method is more costly but saves the expense of "hard tooling" and lead time during the start-up phase of the product.

Double conductor layer flex circuit

Two copper layers are laminated to a base material. Holes are pierced and copper plated to provide interconnection from one circuit side to the other. The circuit pattern is imaged and chemically etched similar to the single conductor layer circuit process described. Cover layers are laminated over both copper sides, each prepunched or machined to clear the unique attachment sites. After final lamination, the circuit is die cut into the specified shape or configuration for assembly processing.

Finish coat option

Following the basic fabrication process steps, the surface mount land patterns and other solder attachment areas can be coated with benzo-triazole, a material that seals the bare copper surfaces from contamination and oxidation or pretinned with solder alloy. If solder alloy is preferred, specify 60/40 tin-lead, but allow for process variables, generally 1 to 2 mils for overall thickness, depending on the technique used to apply the alloy.

Post plating and die cutting

Soldering leads to bare copper contact pads and holes is a common practice. However, tin-lead plating SMT contact locations is preferred. The tin-lead alloy is a more process-friendly medium for reflow-solder attachment of surface mount devices. To provide a more solderable surface to the flex circuit, electroplating or re-flow plating of the tin-lead alloy can be used.

A popular method of solder coating the SMT contact areas is the screen or stencil transfer of a 37/64 ratio tin-lead solder paste to the pattern. The solder paste is then fused to the copper area using a high-temperature reflow followed by a leveling process as shown in Fig. 9-19.

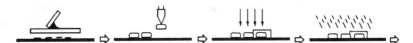

■ **9-19** *The SMT assembly process for the flexible substrate starts with the application of a solder paste on the device land patterns, placement of the device into the solder, and exposure to heat adequate to melt or reflow the solder, followed by a cleaning cycle to remove the flux and other residue from the finished circuit.*

Because of the many variables in this process, some variance in the finished plating thickness should be expected, (Fig. 9-20). The finished thickness tolerance of the solder on the contact areas is difficult to control with reflow-plating technology.

199

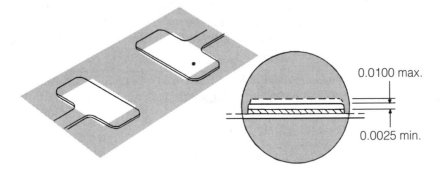

0.0100 max.

0.0025 min.

■ **9-20** *The finished thickness tolerance of the solder on the contact areas is difficult to control with reflow-plating technology.*

Die cutting

To separate each circuit from the panel, the panels are passed through a blanking die designed to "punch-out" the finished profile of each flexible unit. A blanking die can be a relatively low-cost "steel-rule" type fixture for lower volume and short-run products. The hardened-tool-steel, matched-die systems can be developed for the mature long-running parts. *Hard tooling* is more expensive but requires very little maintenance over the life of the product. For flexible circuits with secondary SMT assembly processes, the blanking procedure could be postponed until all assembly operations are complete.

Finish coat or plating options

Following the basic fabrication process steps, the surface mount land patterns and other solder attachment areas can be coated with benzo-triazole, a material that seals the bare copper surfaces from contamination and oxidation or pretinned with solder alloy. If solder alloy is preferred, specify 60/40 tin-lead but allow for process variables, generally 1 to 2 mils for overall thickness, depending on the technique used to apply the alloy.

Rigid-flex circuit fabrication

When the rigid portion of the substrate includes circuit layers, the interface between the rigid and flexible material should be designed as a balanced structure. Ideally the etched flexible portion of the composite is laminated or sandwiched between equal thicknesses of rigid layers. The rigid material might be one or more copper layers and must be fabricated from materials compatible with the base material. The final lamination of the flex circuit to the rigid circuit section is similar to the methods used to fabricate multilayer PC boards with final drilling and copper plating of holes to complete the interface of circuit layers as shown in Fig. 9-21.

Via holes related only to flexible layers of the rigid flex board are processed before final lamination. Common via holes to both the rigid and flexible layers, typical of that shown in Fig. 9-22, are drilled and plated after lamination.

Complex rigid boards are often combined with flexible materials to provide a tough, reliable interface to other rigid circuits or connectors. To provide a reliable product and maximize yield in the manufacture of the circuit, a balanced construction is recommended for multilayer printed circuits. As with one and two layer flex circuits, a cover layer of the polyimide base material is laminated over the copper center sections to protect the flexible section while conventional photo-imaged solder mask is applied to the rigid sections of the circuit.

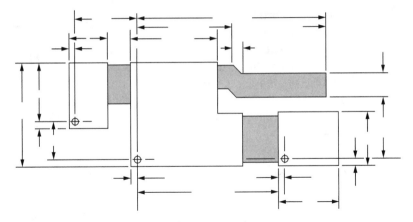

■ **9-21** *When dimensioning flexible circuits its not uncommon to use several datum reference locations to ensure tolerance accuracy in specific or isolated zones of the circuit.*

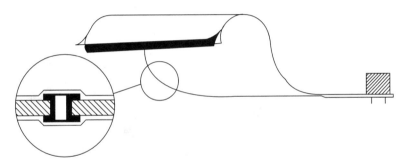

■ **9-22** *Flexible circuits having two metal layers can accommodate plated holes typical of those common in rigid PCB construction.*

High-tech materials for military applications

Ceramic (alumina) materials have traditionally been used in military or extreme environmental applications. But the need for larger substrate size and the less-than-uniform shapes of today's electronic applications have made it necessary to explore alternative materials. Metal core laminated construction of circuit boards using relatively new materials is now being accepted for military and space applications.

A metal core (Copper clad Invar/Polyimide-glass or Polyimide/ Kevlar) laminated substrate is being offered as an alternative material for applications using leadless ceramic chip carriers. Previously, the choice was limited to ceramic substrates in order to match the component's TCE (Thermal Coefficient of Expansion).

The metal core laminate material has been approved for many applications, and companies are specifying CCI in new products being developed. However, extensive testing, technical papers, and reports have not resulted in either military specifications or applicable guidelines to assist the engineer and PC designer in implementing this process. In this next section, practical guidelines are presented for the PC board designer to ensure the most producible finished product.

Specifying Copper-clad Invar

In order to have a clear understanding of CCI fabrication guidelines, it is important to know the process steps. One method of adapting these more stable materials into your products is by laminating the finished PC boards to a layer of CCI to provide stability, ground plane, and a thermal transfer medium as shown in Fig. 9-23.

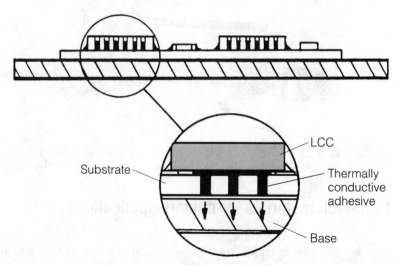

■ **9-23** *Thermal compounds will transfer heat away from the component body and into the metal core layers or base of the substrate.*

The most basic metal core laminated substrate would have one or two CCI layers and one single-sided copper-clad polyimide sheet laminated to both outer layers detailed in Fig. 9-24.

Lamination would take place after clearance holes are drilled or etched in the CCI layers. This will prevent unwanted signal connection when holes are plated through. Clearance holes must allow for insulation around via holes that are not to be connected

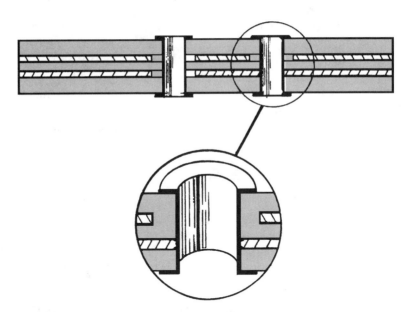

■ **9-24** *Plating via holes from the outside circuit traces to a specific layer of a copper-clad invar core will require predrilling of clearance holes in other CCI layers before lamination.*

with CCI layers. The laminating adhesive film will flow into the clearance holes in the CCI to provide the dielectric separation of copper-plated via holes. After lamination, add all holes to substrate, including those that will connect to CCI layers. The entire substrate panel is plated with one ounce of copper, and simultaneously, holes are copper-plated from one side to the other. Only holes not drilled for insulated clearance will connect to the internal layer, providing ground or plane connection. The circuit image is now plated to the outside surfaces and through the holes. This plating will act as a resist while exposed bare copper is etched away. If a tin-lead alloy is used as the plating medium, the etched panel is generally passed through a reflow process to fuse the tin-lead with the copper, leaving a bright finish.

A solder mask coating can be applied to both surfaces of the finished substrate. Contact areas and via pads should be free of this mask material, as noted for conventional PC boards. The advantages of coating the substrate are: the solder paste used to attach the SMT components is contained on the land pattern, solder bridging is reduced or eliminated, and the bare laminate and conductors are covered, reducing absorption of moisture and contamination.

If space permits, reference designators and component outlines are screen printed on the board surface in epoxy ink. This aids in identifying each device as well as in identifying the polarity and orientation of SMDs. The fabrication procedure will vary from complex applications that require the stability of a metal core laminated substrate. This variable depends on the number of layers required to interconnect the circuit. Surface mount components often are attached to both outside surfaces of the substrate. It might be necessary to reduce the number of side-to-side plated through-holes to avoid excessive perforation of the metal core material.

Many of the circuit interconnects can be made within the layer associated with each respective side. Only the via holes necessary to connect side one with side two and the power/ground CCI layers will be drilled during this step. After the final drilling, the holes are copper plated to complete the connection between layers.

The process requiring blind via holes that do not continue through all layers will add to the circuit board fabrication cost. The continuity of the copper-clad invar layers, however, will be maximized by reducing the number of holes in the CCI. This maintains the continuous core layer, which is desirable to ensure the most thermally stable finished substrate (Fig. 9-25).

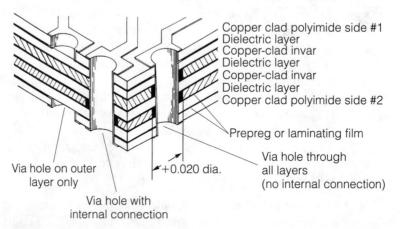

Copper clad polyimide side #1
Dielectric layer
Copper-clad invar
Dielectric layer
Copper-clad invar
Dielectric layer
Copper clad polyimide side #2

Prepreg or laminating film

Via hole through all layers (no internal connection)

+0.020 dia.

Via hole on outer layer only

Via hole with internal connection

■ **9-25** *Predrilling and construction of a CCI core board will require a great deal of planning and interactive communication with the fabrication specialist.*

Most design rules recommended for surface mount technology on conventional epoxy glass substrates will apply to copper-clad invar-polyimide substrates. The assembly process control and solder selection might be more critical, but the components will be attached with reflow-solder technology in the same procedure as normally used in SMT assembly.

Materials for Copper-clad Invar-Polyimide substrate fabrication

Standard material thickness for polyimide-glass and polyimide-kevlar is 0.13, 0.15, 0.20, and 0.25 mm (0.005, 0.006, 0.008, and 0.010 in). Polyimide-Kevlar material presently costs four times as much as polyimide-glass. Check your supplier to learn the standard sheet sizes available. In addition to the basic dielectric, choose the copper thickness desired:

0.015 mm (0.0007 in) thick ½-oz. Copper-clad

0.03 mm (0.0014 in) thick-1 oz. Copper-clad

0.07 mm (0.0028 in) thick-2 oz. Copper-clad

Dielectric materials are available without copper, copper-clad on one side; or copper-clad on two sides, and each might have a different thickness.

Prepreg, or laminating film is used to bind layers together and is furnished in a standard 0.0025-in thickness. This material can be layered to build up to the final thickness specified for the finished board. Prepreg is supplied to the PC board fabricator in 38-in-wide rolls.

Copper-clad Invar material is available in several thicknesses: 0.15 mm (0.006 in) is furnished in rolls. CCI materials that are 0.25, 0.5, 0.8, 1.27, and 1.5 mm (0.010, 0.020, 0.030, 0.050, and 0.060-in) thick are supplied in sheets. Use standard material sizes whenever possible to keep costs under control. However, all of the laminated materials noted are more expensive than conventional FR-4 epoxy glass laminates, but the additional cost of higher grade laminate material is far less than multilayer ceramic-substrate materials, and mechanically the assembly will prove more durable.

Recommendations to ensure successful CCI fabrication

1. Symmetrical printed circuit board laments will furnish equal stresses on both sides of the substrate. The finished PC board should remain flat upon thermal cycling. Bonding the etched copper circuit polyimide layers on both sides of the

205

copper-clad invar also eliminates assembly flexing problems during thermal cycling.

2. A conductor width of 0.2 mm (0.008 in) is preferred. 0.15 mm (0.006 in) is the minimum conductor width on outside layers.

3. Total board thickness range is 0.062 to 0.100 in. with a maximum of 1 inch.

4. Board thickness tolerance is 10 percent of nominal or 0.18 mm (0.007 in), whichever is greater.

5. A 0.20 mm (0.008 in) thickness is preferred for the dielectric between conductive layers. The dielectric layer should not be less than 0.9 mm (0.0035 in).

6. An air gap of 0.2 mm (0.008 in) between conductors is preferred, with a minimum of 0.13 mm (0.005 in).

7. Finished board thickness to plated through-hole size ratios: 3:1 is preferred; 4:1 is maximum.

8. Plated through-hole diameter to finished board thickness of 1.5 mm (0.062 in) inch or less: 0.5 mm (0.020 in) hole diameter. For boards that require 0.075-in thickness or less: 0.63 mm (0.025 in) hole diameter. 2.54 mm (0.100 in) thickness or less specify 0.9 mm (0.035 in) hole diameter.

9. Annular plating around finished hole is 0.18–0.25 mm (0.007/0.010 in). Minimum of 0.13 mm (0.005 in) is to be avoided.

10. Conductor clearance to edge of board: internal layer of 0.100 in. is preferred, 0.031 in minimum. External layer of 2.5 mm (0.100 in) minimum.

Most of these recommendations are applicable to conventional PC board fabrication as well. Specifying standard materials will ensure greater producibility, because this allows the fabricator to use off-the-shelf materials. Overall thickness can be specified, including plating, but do not specify spacing between the layers created by the prepreg or core thickness. If you must specify core and/or prepreg thickness when spacing is critical, give a loose tolerance on the overall thickness for maximum fabrication efficiency and yield.

SMT assembly process

Planning the SMT assembly process

THE OPERATION OF PLACING SURFACE MOUNT DEVICES onto a designated land pattern set or array is referred to as *pick-and-place*. For low-volume assembly or prototypes, hand placement with tweezers or vacuum pickup tools might prove adequate. As component density increases or assembly volume grows, the use of automation for this task becomes more practical.

Speed, accuracy, and versatility for surface mount placement will vary from one equipment manufacturer to another. Factors that must be addressed in choosing equipment for a factory are:

☐ Maximum substrate or board size and placement area.

☐ Volume projected for the product.

☐ Component placements per hour.

☐ Accuracy of placement and repeatability.

☐ Component part mix (passive vs. active).

☐ Number of component stations.

☐ Component packaging (types and sizes).

After evaluating the product and component mix, it might be practical to combine several specialized systems in line to efficiently process the assembly.

Surface mount components are attached to land patterns etched into the circuit. The assembly sequence starts with applying solder paste onto these patterns using a stencil fixture or other dispensing system. The components are placed into the solder-paste-coated land pattern, and reflow soldered. The in-line conveyor-type solder reflow systems are in wide use, with infrared or forced hot air/gas convection through a series of chambers raising the temperature of the assembly until the solder paste is converted into a liquid, reflowing the solder. During this heating process, any solvent or moisture still present from the paste or the board itself is extracted. As the solder cools and returns to a solid state, the electrical and mechanical bond between the component and the PC board is completed, as shown in Fig. 10-1.

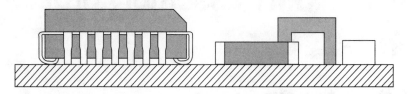

■ **10-1** *The most basic SMT assembly will require only one solder process.*

Double-sided surface mount

When surface mount components are mounted on both sides, as shown in Fig. 10-2, the assembly sequence would first apply solder paste to the land patterns on side one of the PC board.

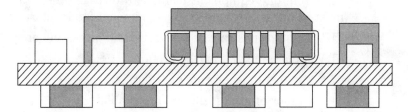

■ **10-2** *Type-2 assemblies require a second solder process. The second process might use wave solder technology.*

The secondary side of the board can be assembled using the reflow-solder process, or as an alternative, components can be attached to the board surface with adhesive epoxy and wave soldered. In this process, wave solder technology will electrically attach the SMT devices on the secondary surface of the board and the leads of the PIH devices attached from the primary side of the board simultaneously as shown in Fig. 10-3.

The leaded PIH components might include conventional DIP and SIP packages, axial-lead resistors, diodes, capacitors, and jumpers or radial-lead devices.

Secondary side wave solder assembly is ideal for attachment of chip resistors and capacitors, SOT and SO packages, diodes, etc. Tantalum capacitors in SMT packages might not always be suitable

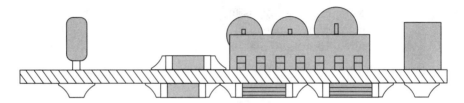

■ **10-3** *The more complex assembly may have mixed technology with surface mount devices on both primary and secondary sides of the board.*

for secondary side mounting because of the higher profile of the component. Wave-soldering of the PLCC also is impractical because of its high profile and lead configuration.

Wave solder for SMT assembly

Wave soldering was the very first approach to adapting SMT to high-volume production of PC boards. The economic advantage is two fold: greater component density and utilization of common wave-solder equipment.

When merited, the designer might use both primary and secondary sides of the PC board for surface mount attachment. The goal of component specialists should be to select as many surface mount alternatives to leaded devices as possible. With few exceptions, most PIH devices are furnished in alternative surface mount packages as shown in Fig. 10-4.

■ **10-4** *IC package styles have been developed to adapt to any complexity of silicon. The SO IC is most commonly used for standard logic and analog devices. The PLCC and QFP packages will accommodate more complex or application specific requirements.*

Two-sided assembly is far more complex than single-sided assembly, and mixing leaded devices with surface mount devices requires thorough planning and possibly additional fixtures for processing. To reduce assembly labor and complexity, restrict leaded devices to the primary side of the PC board.

Reflow process options

Solder methods for reflow include: ovens and induction heating; infrared (IR); conveyorized hot panel; vapor-phase and hot air/gas convection.

☐ A stationary hot-air oven or electrically heated convection furnace, with or without inert gas, can be used effectively to reflow solder paste.

☐ Infrared belt furnaces, with two, three, or more heat zones above and below the assembly provide efficient reflow.

☐ An electric-controlled hot panel, or a series of hot panels, with a belt to transport the assembly is an effective method of reflow soldering the single-sided assembly.

☐ Vapor-phase or condensation-soldering offers precise temperature control.

☐ Conveyor-type in-line hot air/gas convection systems are widely used in high-volume production because of its efficient and consistent heating capability.

Although forced hot air or gas convection is favored by many SMT assembly process specialists for reflow solder applications, infrared reflow soldering systems are also in wide use throughout the industry. Absorption of the IR energy varies with materials used. For example, the organic components of solder paste are excellent absorbers of IR, while gold or aluminum, and previously reflowed solder, are good reflectors. The use of focused or nonfocused IR is an efficient method of selecting reflow soldering for individual components as well. Laser or soft beam soldering with a microprocessor-controlled stepping system is extremely fast; however, the paste might require drying due to the rapid heating cycle of this method.

When vapor-phase is used for solder processing, the entire assembly is heated to the temperature of the vapor of a chemical held at its boiling point (Fig. 10-5). Both batch and in-line vapor-phase reflow systems are available. The batch systems have proved to be very efficient for the lower volume SMT assembly applications,

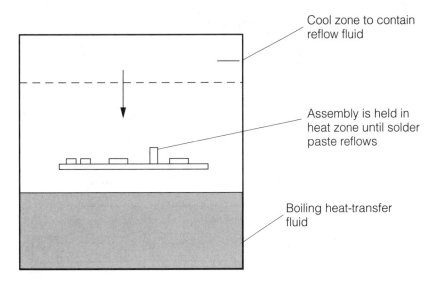

Cool zone to contain
reflow fluid

Assembly is held in
heat zone until solder
paste reflows

Boiling heat-transfer
fluid

■ **10-5** *Vapor-phase solder-reflow methods are ideally suited to those substrates that have a solid metal core, or assemblies with very high component density.*

while the in-line process is better suited to a higher volume manufacturing process.

Cleaning reflow-soldered assemblies

Although several suppliers of attachment materials can supply solder with fluxes that do not require a cleaning process, many product categories might demand some form of flux removal following the assembly operations. The equipment selected for either chemical or aqueous cleaning is dependent on volume and company specifications for measuring assembly cleanliness. Fluorocarbon-type solvents with added polar solvent, for example, have excellent penetration characteristics, but have been banned for commercial applications. Ultrasonic systems using less hazardous solvents are also effective.

A very common method for cleaning SMT assemblies after reflow-solder processing is water. Typically a high velocity water spray cycle is effective in removing organic flux residue as well as foreign particles. In this process residues from water-soluble flux systems should be removed with agitated or heated water, followed by deionized water rinses and forced-air drying. Cleaning options are defined further in chapter 11.

Equipment planning for SMT assembly

The assembly process for mixed technology, Pin-In-Hole (PIH), and Surface Mount Devices (SMT), on the same PC board, requires a close review and understanding of assembly equipment and solder process capability. Figure 10-6 describes a general flow diagram of a flexible medium- to high-volume assembly line for mixed technology PC boards.

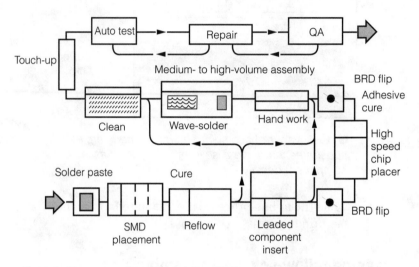

■ **10-6** *The assembly flow of the SMT product can include several process steps. Each process step is audited to maintain the established quality yield.*

The combination of compatible systems are matched to the specific volume and component type mix of the product. Unlike a dedicated line for one product, a flexible combination of systems for specific applications can be set up for a multitude of assembly types and quantity. As a typical example, assuming that the design of the assembly provides for all components to be mounted on one surface, the assembly sequence would be quite simple.

1. Solder paste is first applied to the SMT contact patterns employing a chemically etched stencil fixture.

2. Then surface mount component parts are placed into the solder by a combination of robotic systems.

3. Reflow of solder is achieved by exposing the entire assembly to a high-temperature (190° – 220°C) environment.

The solder paste liquification in the reflow or heating process will complete the mass termination of the components to the substrate.

Alternative solder processes

Heating the solder paste material momentarily converts the alloy to a liquid. The alloy will provide attachment (or wetting) by joining the device lead or contact alloy to the alloy furnished on the PC board land pattern.

Very few assemblies are pure SMT. On the contrary, most have leaded device requirements, and often surface mount components are to be mounted on both sides of the substrate. Here are five solder attachment process examples to consider for surface mount assembly:

☐ Wave Solder (IPC Type 2C)

☐ Reflow Solder (IPC Type 1B)

☐ Reflow-Wave (IPC Type 1C)

☐ Double Reflow (IPC Type 2B)

☐ Double Reflow-Selective Wave (IPC Type 2C Complex)

When the ratio of leaded part to SMT is relatively small, using automation for connectors and headers might be impractical. Often the leaded parts are installed by hand and attachment is completed using a wave solder system.

Planning the assembly equipment and soldering processes to be used on the product requires an understanding of each manufacturing process step. The following examples define the process for five surface mount assembly types:

☐ *IPC Type 1B (SMT only, on one side)* Apply solder paste, place SMT devices in paste, and reflow solder.

☐ *IPC Type 1C (Leaded components and SMT on primary side)* Apply solder paste, place SMT devices in paste, and reflow solder. Install PIH devices and wave solder.

☐ *IPC Type 2B (SMT on two sides)* Apply solder paste, place components, and reflow solder one side repeating the process for the other.

☐ *IPC Type 2C (PIH on primary side, SMT on secondary side)* Install and crimp leads of PIH devices, flip board and attach SMT with epoxy, flip board and wave solder.

☐ *IPC Type 2C Complex (SMT on two sides with leaded components)* Apply paste and attach SMT devices to secondary side and reflow solder. Apply paste and attach SMT devices on primary side and reflow again. Install PIH devices on primary side and wave solder using selective masking fixture.

An understanding of the materials and process options enables the engineer to prepare the Manufacturing Process Instructions (MPI) for the product. The MPI format defines each assembly step and specifies the solder materials, type and model of equipment to be used, temperature of solder operations, and all the parameters required to provide maximum quality and minimize soldering defects. The engineer begins the planning by establishing each of the basic process steps required for the assembly.

Planning SMT equipment

Setting up an SMT production line in-house is frequently the goal of those adapting the technology. Manufacturing services also are used as a line parallel to a company's own operation, thereby building in a safety factor and allowing for reaction to sudden surges in product demand. To plan properly for SMT equipment needs, review the current requirement and future projections for the product

☐ Component types and quantity per assembly.

☐ Number of different assemblies to be processed.

☐ Current and projected volume requirement per assembly type.

Planning the SMT assembly process

The flow of product through a typical assembly operation requires planning. Experienced planners and manufacturing engineers can ensure the best utilization of equipment as well as human resources. Good material planning will ensure that all components are available when needed and these components are furnished in the correct packaging required for the automated SMT assembly systems.

Tape-and-reel packaging is specified for the smaller or passive components, as shown in Fig. 10-7, while ICs are available in both tube magazines and tape-and-reel for the high-volume applications. Because of the delicate nature of fine-pitch devices, tray carriers are generally specified. Communication, coordination,

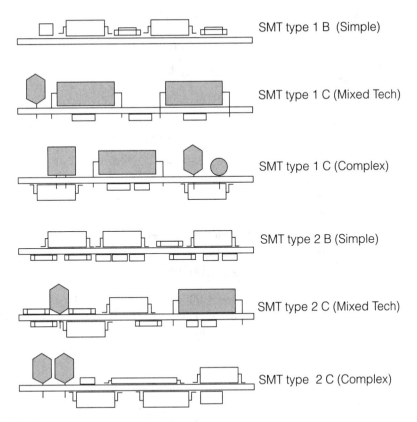

SMT type 1 B (Simple)

SMT type 1 C (Mixed Tech)

SMT type 1 C (Complex)

SMT type 2 B (Simple)

SMT type 2 C (Mixed Tech)

SMT type 2 C (Complex)

■ **10-7** *SMT assembly types.*

and cooperation between the design engineers, buyers, and manufacturing engineers in the planning stage is vital!

Case study product description

The following is a description of an SMT assembly with a small number of leaded devices.

PC board size: 88.9 mm × 284.5 mm (3.5 in × 11.2 in)

SMT devices: 73- Passive components

               24- Small outline ICs

               12- J-lead PLCC ICs

               2- 120 lead QFP ICs

PIH devices: 2- 40 lead DIP ICs

2- PIH connectors*

4- Radial lead devices*

*Hand installed as a post process.

Automated Assembly Process Sequence   This assembly provides for all components to be mounted on one surface.

1. Solder paste is applied to the SMT contact patterns with an etched stencil fixture.
2. Surface mount component parts are placed into the solder by a combination of robotic systems.
3. Expose the partially assembled board to an infrared (IR) or hot air/gas reflow process.
4. Clean and inspect assembly.

Assume that the leaded parts will be installed and soldered in a wave system or by hand. The basic SMT assembly line for reflow-solder technology might include one or more placement machines (see Fig. 10-8), depending on device type and quantity.

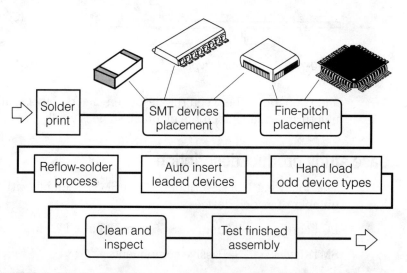

■ **10-8** *The basic SMT assembly line for reflow-solder technology.*

Assembly considerations

To ensure that all assembly process steps meet expectations, all aspects of materials and processes must be defined by a uniform procedure and typically each phase is periodically monitored or audited.

1. Incoming material inspection.
2. Kitting of materials.
3. Baking of PWBs and plastic packaged components.
4. Screen printing of solder paste onto the PWB.
5. Component placement on primary side.
6. Inspection.
7. Reflow.
8. Assembly cleaning.
9. Inspection.
10. Adhesive dispensing.
11. Component placement on secondary side.
12. Inspection.
13. Curing of adhesive.
14. Auto insertion of through-hole components.
15. Hand insertion of odd-shaped through-hole components.
16. Wave solder.
17. Cleaning.
18. Inspection.
19. Test.

To ensure efficient resource utilization, a combination of compatible systems is matched to the specific volume and component type or mix for each product. Unlike a dedicated line developed for only one product, a flexible combination of systems for various applications can be quickly set up for a multitude of product types and quantities.

The following describes the elements that must be considered during the preproduction planning:

☐ Solder material selection.
☐ Solder application method.
☐ Assembly machine utilization.
☐ Solder reflow process.
☐ Cleaning or cleanliness requirement.
☐ Test and Inspection.

217

Solders for component attachment

Solder, an alloy made principally of tin and lead, provides reliable electrical connections and mechanically strong joints. Other metals such as antimony, silver, cadmium, indium, and bismuth are alloyed with tin and lead to control certain physical and mechanical properties of the alloy, e.g., melting range, tensile and shear, and even corrosion resistance.

Tin-lead or "soft" solder alloys are the most widely used in electronic applications, because their low melting temperatures make them ideal for rapid joining of most metals by conventional heating methods. Take care in specifying the proper alloy for each soldering method because each alloy has unique properties. When referring to tin-lead alloys, tin is customarily listed first. For example, 60/40 refers to 60 percent tin, 40 percent lead, by weight.

General-purpose solders include 40/60 and 50/50, which are typically used for plumbing and sheet-metal as well as for high-temperature electrical applications. Where minimum heat must be used during formation of the solder joint, as in surface mount assemblies (with heat-sensitive components and materials), higher tin-content alloys are required, such as 60/40 or 63/37.

Alloys of tin-lead with a small percentage of silver (63/35/2) are used to reduce the leaching of silver from silver alloy end termination of some passive components. These types of alloys are also ideal for soldering to thick-film silver alloy coatings on ceramic hybrid circuits. Bismuth-containing solders, which are frequently used as fusible alloys, can be used in applications where the soldering temperature must be below 183°C (38°F). Indium alloys, also with low-temperature melting ranges, are very ductile and are therefore suitable for joining metals with greatly different coefficients of thermal expansion.

Using solder pastes with SMT

Solder pastes are homogeneous mixtures of a paste-flux and fine-powder solder alloy. The physical and chemical characteristics of the material can be matched precisely to the solder joint requirements, e.g., the method of placement used and required definition, in-process conditions, solder-reflow method used, and cleaning requirements. Because all the ingredients required to successfully place and solder the components are contained in the paste, it is an ideal material for automated assembly of both simple or complex mechanical, electrical, and electronic systems. With refined process development and controlled assembly procedures, reliable solder joints can be repetitively produced.

For the more difficult assembly applications, solder paste might provide the only practical method of solder attachment. Electronic grade solder pastes are manufactured to meet the critical requirements of electronic component assembly. The composition of the pastes can vary with individual requirements. A wide variety of compositions can be specified from suppliers, which comply with recognized standards, like those of the ASTM (American Society for Testing and Materials). Several solder attachment alloys are available for device attachment (see Table 10-1); each one has a unique melting point and can furnish the physical characteristic needed for the specific product or application.

■ Table 10-1 Solder alloy for device attachment.

High temp. alloy	Melting temp. C
96.5Sn/3.5Ag	221–226
95Sn/5Pb	222–227
96Sn/4Ag	238–243
95Sn/3.5Ag/1.5In	218–223
Standard alloy	**Melting temp. C**
63Sn/37Pb	185–190
62Sn/36Pb/2Ag	179–186
62Sn/36Pb/2In	179–185

In addition to selecting specific alloy compositions, the engineer must consider flux type required for the process. Both organic and inorganic fluxes are available and cannot be mixed in the various assembly processes. Organic systems might be preferred for ease of water cleaning and are generally environmentally acceptable. Organic solder-flux residues are very corrosive and must be removed from the board surface after reflow.

An example of a no-clean flux is the RMA. Mixed into the solder, it provides a paste-like material consisting of the following primary components:

☐ Solvent or vehicle.

☐ Rosin/resin or organic.

☐ Activator.

☐ Viscosity-control additives.

Solvent selection is based on compatibility and ability to dissolve high concentrations of solids. Other important properties are:

☐ Slow evaporation at room temperature, low moisture absorption, high flash point, and compatibility with supplemental activators and viscosity modifiers.

☐ The solids used in solder pastes are selected for their unique characteristics. These characteristics might include: cleaning of surfaces to be soldered, oxidation protection to the solder powder and solder joint during heating, binder and surface protection after curing, stability at soldering temperatures, and if residues are still present after reflow processing, the ability to be completely removed by conventional cleaning methods.

☐ Activation levels of flux include: nonactivated (R), mildly activated (RMA), activated (RA), and super-activated (RSA).

☐ Organic thickeners are added to flux systems to alter physical properties as required for typical applications. Viscosity is selected based on methods of dispensing—screening, stencil printing, or others.

Solder application

Solder pastes are applied to the PC board land patterns in several ways, including syringes and pressure-fed reservoirs. Large or small amounts of paste can be dispensed this way providing single dots or strips that can be controlled with timing devices or robotic applicators. A more common method of solder paste application uses chemically etched solder stencils. This technique of depositing solder paste is widely used for medium- to high-volume assembly.

Using a squeegee, solder paste can be accurately transferred onto a surface, as shown in Fig. 10-9 by a manual or automatic operation through an etched stencil or screen. Mesh size of solder screens (ranging from 80 to 200 mesh) depends on the definition required. Paste thickness and volume is controlled by screen wire and opening size, emulsion, or mask thickness. The squeegee normally employed to force paste through openings in the screen will transfer solder paste material to only the land patterns of each device.

Transferring solder paste with a stencil improves both volume and registration accuracy. The stencil material, usually brass or stainless steel, can be specified in various thicknesses. The surface

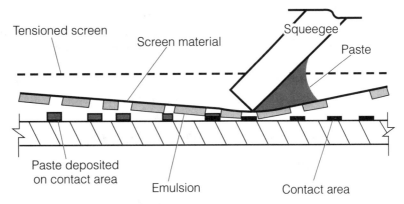

■ **10-9** *Depositing the solder paste to the contact land patterns of the substrate is efficiently accomplished with screen or stencil fixtures.*

mount component land pattern is chemically etched through the material, leaving a precise opening for solder paste transfer. Thickness of the paste is determined by the gauge of the stencil material.

For SMT assemblies with fine-pitch QFP devices, the stencil material is superior to the screen. Multilevel stencils are possible as well (see Fig. 10-10). For example, the process engineer will have the option of depositing 0.15-mm (0.006-in) thick solder paste on the fine-pitch IC land pattern, while maintaining a 0.2 mm (0.008 in) thickness of the alloy for the larger surface mount device type.

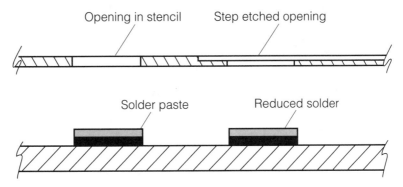

■ **10-10** *Applying solder paste in varying thicknesses at select areas of the substrate can be accomplished with selective etching of the stencil material.*

Applying solder paste in varying thicknesses at select areas of the substrate can be accomplished with selective etching of the stencil material; however, component clearance to the fine-pitch ICs must allow for the squeegee to conform to the recessed area. Consistent solder paste resolution can only be maintained if the squeegee is in full contact with the stencil surface during the transfer sequence.

Providing for automatic vision alignment

The fiducial target method of alignment is used for automatic solder stencil and assembly operations. Three fiducial targets are required on each panel to permit automatic PC board to stencil alignment. The fiducial location can be outside the assembly area of a panel array or on each individual unit. To provide for automatic placement of fine-pitch and TAB devices, the addition of two targets within or near the device contact pattern is advised. (See Fig. 10-11.)

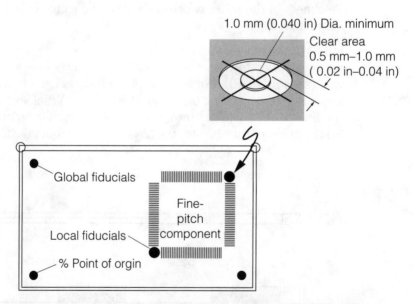

■ **10-11** *To provide for automatic placement of fine-pitch and TAB devices, the addition of two targets (on each corner of the top edge) within or near the device contact pattern is advised.*

Fine pitch usually defines a high pin-count IC with center-to-center lead spacing of 0.63 mm (0.031 in) or less. Many QFP devices now being used for custom and semi-custom applications are using 0.63, 0.5 and 0.4 mm (0.025 in, 0.020 in, and 0.015 in) spacing. The trend moving toward even closer spacing requirements will continue until the limit of circuit board fabrication technology and the placement accuracy of component assembly systems or process yield reach unacceptable limits.

To mount surface mount components on the secondary side for assembly, an adhesive epoxy is applied by dispensing or by a pin-transfer method to retain the device, as in Fig. 10-12 and cured to harden in position. Stencil application of epoxy for component attachment on the wave-solder side is a practical option to dispensing epoxy, especially when using the smaller devices. With the parts bonded in place, the board is flipped back to the primary side and all remaining component leads are inserted through plated holes in the board.

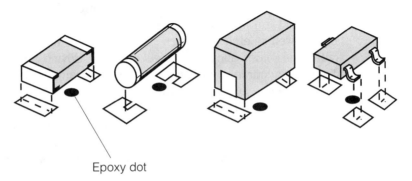

Epoxy dot

■ **10-12** *Discrete components are available in standard EIA packaging that will adapt to most assembly systems.*

If leaded parts are to be machine assembled, the epoxy attachment of SMT components will generally follow insertion and crimping operations. Following the above sequence, the assembly is passed through a process that includes flex application, pre-heating and wave-solder process.

Selecting solder paste for fine pitch

After evaluating several solder types, two or more suppliers might be selected to furnish material by meeting the acceptable criteria of defect free processing. The SN63 (63 tin/37 lead) alloy can be furnished with very fine powder size (–325 + 500 mesh) having 95.5% metal, an ideal material for fine-pitch attachment. Evaluating solder materials and process characteristics involves both comparative testing and process engineering.

Viscosity testing

The viscosity of the solder paste material must remain uniform throughout the process. If allowed to dry out during the stencil process, the stiffened material will not transfer evenly to the board. A viscometer is often used to test solder paste material on a sampling basis for each container opened. For best results, manufacturers recommend viscosity testing of the material at 25°C temperature. With a spindle rotation of 5 RPM, the viscosity recommended must remain in a range near 850 kcps (2600 kcps for the Malcomb System).

Fine-pitch solder stencil development

Based on earlier experience of processing fine-pitch devices in high-volume assembly, the engineering team might rely on several factors that influence the stencil development. The stencil thickness, as well as etched pattern geometry, determines the precise volume of solder alloy deposited onto the device land pattern. Alignment accuracy and consistent volume transfer is critical for uniform reflow-solder processing. The material selected for the stencil fabrication will determine both the chemical milling process method and wall finish quality of the opening. Brass, for example, maintains a very uniform wall finish after etching the open features, but brass is not as durable as stainless steel.

Chemical etching of stainless steel is more difficult and generally more expensive than brass, but the stencil is very durable. Of significant concern is the quality of the openings for fine-pitch land patterns. Wall finish on the opening features of stainless steel are not always uniform due to the limits of the etching process, impacting clean solder transfer to the smaller features required for fine-pitch attachment.

The aspect ratio between the width of the chemically etched opening and the solder stencil can be close to 1:1. For example, a full width opening of 0.20 mm etched pattern into the 7 or 8 mil (0.18- or 0.20-mm) thick sheet material is at the threshold of the chemical milling technology. Figure 10-13 illustrates one technique of overcoming the problems associated with stencil printing with narrow pattern geometry through tapering of the opening using chemical or laser technology.

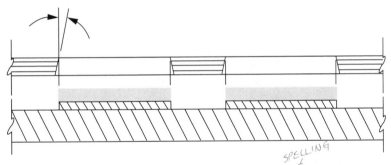

■ 10-13 *Providing a slight taper in the stencil openig* spelling *promotes a very uniform and efficient transfer of the solder paste onto the board surface.*

Alternative stencil materials

Other durable alloys are available for stencil applications and some have more forgiving etching characteristics. One alloy, although not widely used for general stencil requirements is Molybdenum. The material, more expensive than stainless steel and having a rather porous finish, retains a solder paste residue on the surface and requires periodic cleaning. The "Moly" material can be selected for fine pitch because of the decidedly superior wall profile expected. If adapted for production applications, a suitable plating can be applied to the stencil solving the porosity issue.

Even transfer of the solder paste to the board is of primary concern. To reduce dryout of the paste material deposited on the stencil surface, the entire work area of the stencil system should be covered and environmentally controlled. If available a squeegee durometer of 95 or harder should be used. The blade angle and travel speed must also be fine-tuned to ensure even paste transfer. When these elements are not in concert, paste deposit becomes too erratic, especially through the narrow openings parallel to the squeegee blade.

Solder volume measurement

Under normal in-line assembly processing only periodic sampling is customary for process monitoring. During process development, however, it is important to record the solder volume and consistency on each substrate sample.

A Laser Scanning Microscope is often employed to measure height, length, and width of the deposited alloy at several locations on each board. With all measurements recorded, a comparison of possible solder defects due to paste transfer quality can be analyzed and correlated after reflow. To avoid excessive slumping or dryout, the board must be transferred from inspection into the next assembly stage with little delay.

Reflow-solder processing for no-clean solder

Reflow of no-clean solder alloy on fine-pitch assemblies has proven very successful using a forced convection as well as vapor phase system. The convection system should allow for very precise thermal profile adjustment and provide for the optional use of nitrogen, an inert gas, to ensure a very pure reflow environment especially for no-clean solder processing. Vapor phase, on the other hand, occurs in an inert or oxygen-free reflow environment.

Concerns and process refinement

The defect ratio from one assembly to the other might not seem to be affected by the reflow process as much as the solder paste characteristics. The viscosity of the solder paste will have an even more significant impact on the solder joint defect ratio of 0.4-mm pitch devices. Uneven transfer of solder paste will cause random solder defects that will affect process yield as well as product reliability.

Although solder volume for each solder joint can be mathematically modeled, visual inspection on the finer pitch devices might not identify all solder defects. The use of more advanced inspection systems for fine-pitch devices is inevitable. Ultrasonic imaging, X-ray, and X-ray laminography might prove to be very effective in performing a nondestructive solder quality "certification," measuring solder density or detection of solder voids. However, advances in automated electrical testing (Test Jet ICT) will detect assembly process defects as well.

Aqueous cleaning for surface mount assemblies
Elimination of CFC materials from the SMT manufacturing environment

REFINEMENT OF SOLDER PASTE WITH A WATER-SOLUBLE flux has become a primary development effort by everyone in the solder material industry. Although water-soluble flux has been used successfully for years in wave-solder processing, surface mount assembly process specialist favored solvent (CFC) for cleaning.

To comply with the growing concern over long-term atmospheric damage caused by specific chemical vapors of chlorofluorocarbons (CFC), several chemical alternatives have been developed for cleaning electronic assemblies. Although alternative solvent cleaning materials have been proven as a substitute for the CFC chemicals, the high cost of these materials and cleaning systems has proved a deterrent for most commercial applications.

One reason for using the RMA-solvent cleanable-flux material is the extended working time it can provide. The paste can be transferred to the substrate and set for long periods of time before re-flow processing. This is an advantage for small-volume, batch-type assembly operations. On the other hand, most in-line assembly processes do not require an extended time duration because the time between paste application and placement is within minutes. The tack time of the paste material will be affected by any number of environmental variables, especially humidity. Most of these variables can be overcome by controlling the atmosphere within the stencil operation and not allowing the material to stand dormant or exposed for long periods of time.

While chemical alternatives were being developed to remove RMA flux residue, new water-soluble fluxes and water-cleaning systems were also being refined to meet the most stringent cleaning specifications. Water continues to be the most abundant and low-cost method for cleaning, and cleaning agents (saponifiers) have gone through a technical renaissance in recent years, furnishing a reliable and economical way out of the CFC controversy.

The water-soluble fluxes are primarily an organic acid base and are compatible with either wave-solder applications or they are blended with solder paste for reflow processing. Because the cleaning systems and materials are so very different, the shift from chemical to water clean should be evaluated with care.

Of course, not all electronic products (space gadgets, electronic implants, etc.) can rely on water as a cleaning agent, but a vast majority of electronic products can. The main criteria for how effective the cleaning system performs will be based on whether or not the assembly can pass the measurement criteria established by individual companies.

Although cleaning processes are outside the responsibility of the PC board designer, the designer must be familiar with cleaning options because the substrate design and component spacing plays a significant role in how effective the cleaning process will be. Large ASIC devices, for example, might have a minimum spacing that restricts cleaning efficiency. If overall height restriction is not a factor, specify a 0.25 mm to 0.40 mm (0.010 in to 0.015 in) standoff height on devices. The high surface tension of the water actually enhances the cleaning action and in conjunction with high-pressure sprayers, the cleanliness of the assemblies can compare to, or exceed, the quality level of solvent washing. It also has been demonstrated that effective cleaning can be achieved between the larger SMT device body and board surface when minimum clearances, as shown in Fig. 11-1, can be provided.

Cleaning under QFP devices might not be possible because of the minimal clearance between the component and board. Trapped flux under devices can cause damage to plated holes, but by tenting over concealed via pads, the plated holes will not be affected.

Some products going into low-stress environments might not require cleaning at all. No-clean solder fluxes are now available that virtually vaporize in a hot-air or nitrogen atmosphere reflow process. The main issue, for those presently using CFCs as a cleaning vehicle, is that their days are numbered. In addition to a worldwide agreement to reduce or eliminate the use of CFC material in cleaning operations altogether, some cities and countries have banned its use without notice.

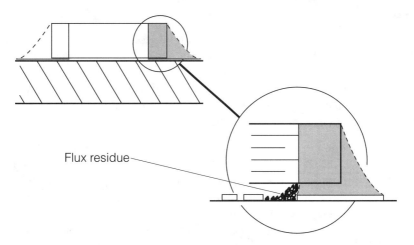

■ **11-1** *After reflow-solder processing of the assembly, a high-velocity cleaning cycle is incorporated to remove all flux residue and foreign particles.*

Identify impact on materials and process

Several issues must be addressed when converting from a solvent to a water-clean process that can impact product quality or long-term reliability:

- [ ] PC board layout and component density.
- [ ] Circuit materials and component lead design.
- [ ] Solder paste selection and flux type.
- [ ] Cleaning system evaluation.

Each of these key factors must be considered when converting to water as a primary cleaning agent. Switching to a water process must never compromise or jeopardize the quality or intended operation of the product.

Of course, all existing designs might not meet all cleaning criteria. For some products, the component density and surface clearance restricts water cleaning altogether, but new designs can anticipate cleaning system limitations by providing adequate space and clearance or adapt the no-clean solder material.

PC board layout and component density

A majority of the flux residue and particles associated with reflow-solder processing will be confined to the area in and around the device leads. Of primary concern are the corrosive contaminants of organic fluxes that can be trapped under the body of the component. These corrosives might not affect the performance of a product initially, but long-term reliability will be at risk. When organic residue is allowed to make contact with exposed via pads or alloy features on the PC board surface, the metal features will actually dissolve. Humidity and high temperature in the operating environment will further activate these residues.

For the spray nozzles to have maximum access to flush out particles, the angle of the spray must not be obstructed. SMT component orientation is not a significant factor in cleaning system efficiency, but other factors will influence cleaning quality. Spacing between components and even distribution of mass will affect both reflow-solder processes and cleaning efficiency. As an example, avoid positioning higher profile devices near large, low-profile fine-pitch quad flat packs. Close spacing of tall components like SOJ or PLCC and other high-profile parts is to be avoided as well. Devices having a height greater than 5.0 mm (0.200 in) should be spaced at no less than 2.5 mm (0.100 in). This is especially true for "J" lead PLCC type components. Even at 2.5 mm (0.100 in) spacing, typical of that shown in Fig. 11-2, visual solder joint inspection and solder tool access is restricted.

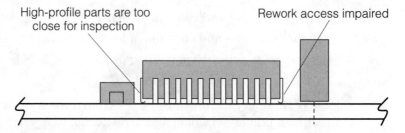

High-profile parts are too close for inspection

Rework access impaired

■ **11-2** *Solder process residue may be more difficult to remove in areas where taller devices are closely spaced.*

Although the water molecules are heated and sprayed at high velocity, the cleaning efficiency on stacked surface mount assemblies might be compromised and a cleaning alternative using a safe level ultrasonic agitation to flush out trapped particles might be necessary.

Solder mask and cleaning

Solder mask coatings are applied to the exterior surface of the finished board and will protect most of the circuit features, but mask must not cover specific locations: surface mount land patterns, via holes and/or pads for testing, and leaded device attachment points. Locating exposed via pads and holes under low-profile quad flat pack and small chip devices should be avoided because of the concern over the efficiency of aqueous cleaners.

Solder mask specified for the majority of surface mount substrates is a *photo-imaged* material. These materials are applied as either a dry film coated wet or a combination of liquid and film. In general, mask coating must be tough and have a low overall profile. One valid concern is that flux or even solder particles can be trapped in the recesses or openings in the mask coating surrounding via pads. See Fig. 11-3 for details.

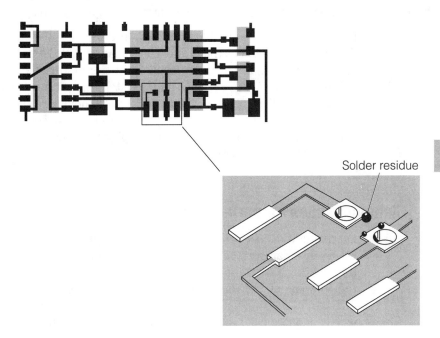

■ **11-3** *Solder residue deposited along the edges of the holes on a PC board.*

To ensure that corrosives are not trapped in the via recesses, solder mask is often used to cover (flood) or tent over the holes under QFP devices. This is an effective way to avoid damage from trapped residues, but those vias should be coated on both primary and secondary surfaces when possible. If access is needed for testing but the via pad is not clear for probe contact, the addition of a satellite pad, clear of solder mask, but without the plated hole, can be employed. See Fig. 11-4.

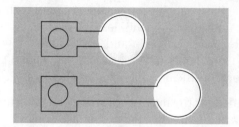

■ **11-4** *If access is needed for testing, but the via pad is not clear for probe contact, a satellite pad, clear of solder mask, but without the plated hole, can be used.*

As noted in chapter 9, dry film solder mask products are generally 0.07-mm to 0.10-mm (0.003-in–0.004-in) thick. This thickness creates a deep pocket in and around the SMT land pattern or via pad and hole restricting cleaning. The thicker mask material might have a negative effect on solder paste stencil definition as well.

High-density circuits and those products using low-profile fine-pitch devices must maintain a thin solder mask coating. At least two solder mask manufacturers have developed a wet transfer/dry film that is only 0.025-mm (0.0015-in) thick for fine-pitch applications. This material has excellent physical properties and will hold very small features, but cost of material and special machines needed for its application must be considered. Liquid photo-imaged solder mask retains a significantly greater acceptance worldwide due to its lower overall cost.

Cleaning issues for surface mount devices

Passive devices seldom pose a cleaning problem because of their relatively small surface area. After soldering these small devices, the clearance or standoff height between the bottom surface of the component and the mounting surface of the substrate will be 0.10 mm to 0.15 mm (0.004 in–0.006 in). This space generally provides adequate clearance for aqueous cleaning systems.

The small-outline (SO) IC has a wider standoff height range from 0.10 mm to 0.30 mm (0.004 in–0.012 in) and plastic leaded chip carrier (PLCC) or "J" leaded devices are designed to allow 0.25 mm (0.010 in) or more air gap between the bottom surface of the device and substrate surface. Assuming the clearance noted, flux residue and particles can easily be flushed away with an aqueous system.

Concerns for unique stacked assembly types

Of particular concern is the cleaning efficiency of the 1 MEG and 4 MEG memory SIMM. It is common practice for the memory components to be spaced side by side with no more than 0.63 mm (0.025 in) separation between leads. In addition, special bypass capacitors are mounted under the memory device body and simultaneously reflow soldered. Details are shown in Fig. 11-5.

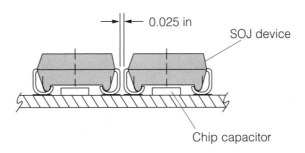

■ **11-5** *Special bypass capacitors are mounted under the memory device body and simultaneously reflow soldered.*

Efficiency on fine-pitch devices

Quad flat pack IC standards are not as uniform as the SO and PLCC family of devices. The *standoff height*, or distance between the bottom surface of the part and the mounting surface, is often open to the discretion of the component manufacturer. Some suppliers, however, are cognizant of the cleaning issues and have provided a reasonable clearance. As an example, the JEDEC-registered plastic quad flat pack (PQFP) shown in Fig. 11-6, has a standoff height of 0.5 mm to 0.0.8 mm (0.020 in–0.030 in) to ensure cleaning of residue from under the device.

The recommendation for lead forming the Carrier Ring and TAB style fine-pitch devices will provide 0.25-mm to 0.33-mm (0.010-in–0.013-in) standoff height. The leads of this type of device are generally formed and trimmed by the end user before, or as a part of, the assembly process.

The EIAJ has established dimensional guidelines for the miniature TSOP and SQFP family of devices that is widely used for products needing a very low overall surface profile. The tolerance limits on these devices are quite liberal, as are most ICs from Japan. Standoff height will probably vary from one supplier to another, but zero clearance is not unusual. By establishing a line of communication

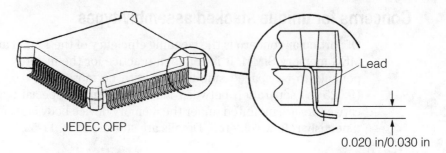

JEDEC QFP

Lead

0.020 in/0.030 in

■ **11-6** *The JEDEC-registered plastic quad flat pack (PQFP) has a standoff-height of 0.5 mm to 0.0.8 mm (0.020 in. to 0.030 in.) to ensure cleaning of residue from under the device.*

to the supplier the end user can specify a clearance of 0.25 mm (0.010 in) to ensure the minimum standoff height needed for reliable water-type cleaning.

Testing solder paste materials

After selecting the specific solder alloy and powder size desired, solder material manufacturers will submit samples for evaluation and comparative testing (on request). Solder materials are in a very competitive market, so most suppliers will respond to requests for samples to evaluate the qualities of their materials. To thoroughly evaluate these materials, a list of key factors should be defined by the process engineer to grade performance of the solder paste. Specific characteristics that must be evaluated might include: viscosity; fixture clean-up; printing characteristics; Omega Meter test solderability; presence of excessive white residue, compatibility of wave solder flux, and how efficient the cleaning after reflow processing.

Example of testing program description

The following is the description of a typical evaluation for solder materials and process:

1. *Viscosity* Each sample of solder paste is checked at incoming Q.A. to determine if the viscosity stated by the vendor was within specification (±10% of viscosity). A typical test can be made using a Brookfield viscosity meter with a spindle rotation of 5 RPM, for two minutes.

2. *Solder print* A stencil is most likely used to print each sample of paste on a number of test board blanks. A sampling

of typical components might include active and passive devices and if the product planned will have the requirements, fine-pitch QFP devices should be furnished.

3. *Reflow process* Assembled boards are typically reflow soldered within a half hour after the solder paste is applied. Board temperature generally will exceed 180°C for 60 to 70 seconds and might exceed 210°C for 20 to 30 seconds during the same period.

4. *Cleaning* Clean boards within two hours after reflow processing.

5. *Cleanliness tests* Ionic residue measurement is typically checked using an Alpha Metals Omega Meter 600R (or similar system), and containment can be detected using surface insulation resistance SIR measurement.

Following the cleaning process, engineers and quality personnel will review assembly samples and examine each solder connection under a microscope to verify solder-joint quality, detect process defects such as flux residue, solder balls, or the presence of residue that might contain active or corrosive materials.

Refer to Table 11-1 as an example of grading the performance of solder paste by one of the leading manufacturers of surface mount assembly technology.

■ Table 11-1 Solder material rating compares the characteristics from various suppliers.

Parameters/vendor	A	B	C	D	E	F	G	H	I	J	K
Viscosity test P/F (score)	F/1	F/1	F/1	F/1	F/1	P/2	F/1	F/1	F/1	F/1	F/1
Screen cleaned with water Y/N (score)	N/1	Y/2	N/0	Y/2	Y/2	Y/2	N/1	Y/2	Y/2	Y/2	Y/2
Printing characteristics P/F (score)	P/2	F/1	F/0	P/2	F/1	P/2	F/1	P/2	F/1	P/2	P/2
Omega meter avg. > 20 Mn-Cm P/F (score)	F/1	P/2	F/0	P/2	F/1	P/2	F/1	P/2	P/2	P/2	P/2
Solderability P/F (score)	F/1	F/1	F/0	P/2	P/2	P/2	F/1	P/2	P/2	P/2	P/2
White residue Y/N (score)	Y/1	N/2	—	N/2	Y/1	Y/1	Y/1	N/2	N/2	N/2	N/2
SIR results Y/N (score)	N/1	N/1	N/0	N/1	N/1	Y/2	N/1	N/1	N/1	Y/2	N/1
Compatible flow solder flux Y/N (score)	N/1	N/1	N/0	N/1	Y/2	N/1	N/1	N/1	N/1	Y/2	Y/2
TOTAL score	9	11	1	13	11	14	8	13	12	15	14

Although specific manufacturers participating in this test cannot be identified by name, the data shown in the table indicates how each sample might be compared during a typical evaluation program. Of the eleven suppliers evaluated in one test, three met the solder quality criteria established. Curiously, excessive residue was seen on only a small sampling of the boards processed. It appeared that the residue was present specifically on the boards using inadequately cured dry film solder mask.

Cleaning systems evaluation

The long-term reliability of a product and its performance can be affected by how well it can be cleaned. The acids or small particles left on the assembly might not be a factor in the initial function of the product. Over time, corrosive action can damage components, the circuit board plating, or even the function of the product. The particles that remain lodged under a component are often small solder balls. These metallic balls might dislodge and roll around, short two signals together, or worse, connect a power supply trace to a ground trace and pooof!

The test methods for evaluation of cleaning systems can include the Bellcore TR-TS4-000078 or MIL-P28809A, a portion of IPC-TM-650, ANSI/IPC-SF-818, or ANSI-JSTD-002.

Cost benefit of aqueous systems

Converting to a water or aqueous cleaning system has a significant benefit related to cost. The operating cost analysis of a CFC solvent cleaning process in a volume-manufacturing operation for example, might reach $250,000.00 a year per system. This includes the solvent material, maintenance, electricity, and in some locales, hazardous waste reclaiming services.

Water too, has costs. The service for supplying deionized water and softeners, saponifiers, filtration media, and utilities are significant, but water cleaning is dramatically less costly than maintaining a CFC cleaning system. As an example, the operation costs for a water-clean system when compared to an equal volume of assembly units processed with CFCs is approximately $6,200.00.

Cost of the in-line aqueous cleaning systems typical of that shown in Fig. 11-7 range in the area of $85,000.00 or more, very similar to the cost of solvent systems. Judging the quality of the cleaning process is a major factor but reliability, factory service, spare parts

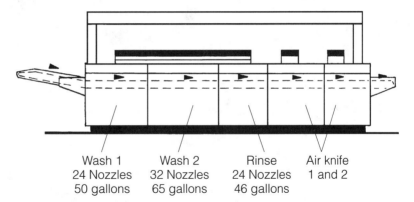

Wash 1　　Wash 2　　Rinse　　Air knife
24 Nozzles　32 Nozzles　24 Nozzles　1 and 2
50 gallons　65 gallons　46 gallons

■ **11-7** *Water cleaning systems have several stages of high velocity sprayers to flush out and remove solder process residues.*

availability (how long will your factory be idle if a part has to come from afar) also must be considered.

Water filtering and disposal

Before the water is disposed of in a municipal waste network the liquid is filtered to remove solid particles. The filtering system must be approved or certified by the local municipality. Most of the cleaning water is recycled as part of the machine's operation. The solids extracted during the filtering process are disposed of by the water service company. Solids that are retained in the water directed to the public waste system require filtering as well.

Although the flux residue is safe and not a threat to the environment, the waste water must be processed through an approved filtering system to capture solids like tin and lead. Filtering systems are approximately $40,000.00, but that is a tolerable expense when compared to the ever-rising cost of replacing and disposing of chemical cleaners.

Extensive testing and evaluation of materials and equipment must be performed by each company's engineering staff. Following testing and a careful review of the results, the engineering team can determine whether proper matching of flux, saponifiers, and cleaning systems has been achieved. The evaluation must prove that through capillary action between the component and substrate, an aqueous solution can flush the flux successfully from the solder joint area.

Design evaluation for efficient assembly processing

A TASK GROUP IS FIRST ASSEMBLED TO EVALUATE THE producibility of an SMT product. The review committee might include representatives of several disciplines including design, manufacturing, and/or process engineering, test, quality, and materials. The review might provide engineering data or identify potential process problems that manifest themselves as additional product costs. The list of concerns might not necessarily be assembly problems or, more specifically, oversights; however, they should be identified because many of them will impact manufacturing efficiency. Concerns fall into several categories including documentation errors, potential interference problems for automatic placement of SMDs, incorrect component land pattern sizes, and poor printed circuit board interconnect techniques.

Before the critique of the surface mount assembly, prepare a checklist of primary concerns that might affect each phase of the assembly process.

Inspection of the PC board

A sample checklist would be similar to the following:

☐ Board flatness and appearance.

☐ Tooling holes for machine handling.

☐ Photo-imaged solder mask (SMOBC).

☐ Solder mask over bare copper.

☐ SMT contacts free of mask.

☐ Screen legend is sharp and readable.

☐ Legend does not overlap pads or SMT contacts.

☐ Trace width and air gap.

☐ Hole size to pad ratio.

☐ Hole registration/breakout.

☐ Plating quality and uniformity.

☐ Efficient use of Material.

☐ Bare board test certification.

Verify dimensional accuracy

Inspect the substrate to the specified IPC quality level established by the design engineer. One solder sample should be furnished with each board lot or date code. Following bare board inspection, apply solder paste to the SMD land patterns on sample of nonpopulated circuit boards. Reflow the solder paste and clean. The solder quality can be evaluated as well as the substrate's physical reaction to the temperature cycles of reflow systems.

If an in-line reflow system is used, one or more thermal couple probes can be attached on the substrate surface to establish basic temperature profiles for the reflow system. These profiles will assist in establishing temperature and conveyer speed during reflow of the populated assembly. Before the board is populated with components, inspect the quality of the plated via and lead holes. Cracks or voids in the plating could be a sign of more serious fabrication problems. If solder shows signs of separation from the land pattern surface, the board is probably contaminated and will not yield a reliable product. The quality and product yield of the surface mount assembly is directly related to the quality of the substrate.

Assembly documentation

Assembly details for surface mount circuit boards must define the location and orientation of all component parts. This will include the component outline and reference designation for each device, including connectors and hardware. If silkscreen legend is furnished on the PC board, an orientation identification mark on the substrate surface should be provided for ICs, diodes, and all polarized devices. Pin one of the IC or connector, as an example, can be defined by a small round dot or other shape on the screened legend as detailed in Fig. 12-1.

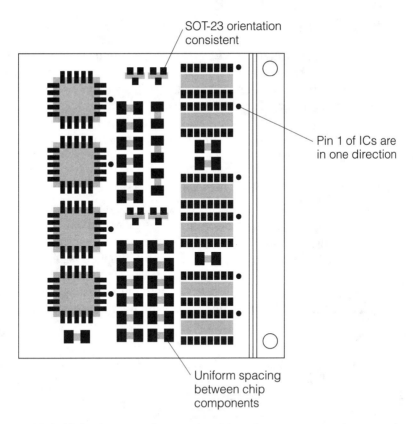

SOT-23 orientation
consistent

Pin 1 of ICs are
in one direction

Uniform spacing
between chip
components

■ **12-1** *Maintain a consistent orientation of components when possible. The direction of components should be clearly defined on the substrate surface and assembly documentation.*

Graphic shapes representing component bodies are often added to the legend master to clarify position on the board surface, but if component density is very high, the designer should consider omitting device outlines. When preparing assembly documentation, provide a composite of the screen legend graphics reference designators and land patterns. Outlines of components, although not furnished on the legend master, can be added to the assembly drawing. Assembly documentation must furnish enough information for processing and inspection of the final configuration. When necessary the final assembly detail will include auxiliary views to clarify mounting of unique components and notes to address special requirements.

Bill of material

The bill of material (BOM) furnishes component description and type, manufacturer's part number, reference designator, and quantity. Data sheets or specifications should be furnished for any components classified as special. An approved vendor list (AVL) also should be included as a controlled part of the documentation and when possible, define two or more approved sources for each part in the BOM.

Fabrication detail

Circuit board fabrication drawings provide base material requirements, physical size, and specific notes describing the finished product. Holes are identified and finished size and tolerance clearly defined. Multilayer substrates will require a cross-section detail, as shown in Fig. 12-2, defining circuit layers and finished thickness. For more information on board fabrication and material options refer to chapter 9.

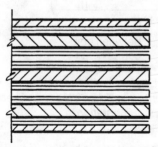

0.005 One side copper cap layer
0.010 Two side copper
0.005 Dielectric layer
0.010 Two side copper
0.005 Dielectric layer
0.010 Two side copper
0.005 One side copper cap layer

■ **12-2** *A cross-sectional view of the multilayer substrate will identify the copper-clad weight, dielectric layers, and finished thickness specification.*

Photo tool (art work) review

Following the documentation review, inspect CAD-generated photo-tool artwork that is used to manufacture the circuit board. Measure component land patterns and compare data to the device specifications. The wrong calculation of contact spacing for a fine-pitch QFP, for instance, will cause extensive hand rework after reflow solder. Inspect clearances between traces and via pads on internal layers. The conductor width and air gap is often far less than the circuit patterns on the board's outer surface. Any clearance less than 0.13 mm (0.005 in) should be avoided, and the same is true for trace width.

Post process assembly evaluation

Following attachment and reflow soldering of surface mount devices, inspect the solder fillet on each component contact. Study the alignment of the passive components and the finished registration of multi-lead IC devices.

Troubleshooting assembly problems requires a close look at cause and effect. First of all, identify and itemize the defect to determine if quality issues are influenced by board design, component quality, process control, or a combination of each.

Solder attachment of a few leaded devices on a two sided surface mount assembly can be performed by hand tools or, using a custom designed holding fixture, selectively wave soldered by machine. In order to use this assembly method, the designer must provide additional spacing between the leaded device and surface mount device previously attached to the secondary side of the board. The additional space will allow for the fabrication of the wave solder fixture. Usually machined from a physically stable dielectric sheet stock, the fixture will provide a barrier or mask, shielding the reflow solder attached devices from additional exposure to the wave solder process. A narrow web of the fixture material is retained around the open area exposing only the selected device leads as illustrated in Fig. 12-3.

243

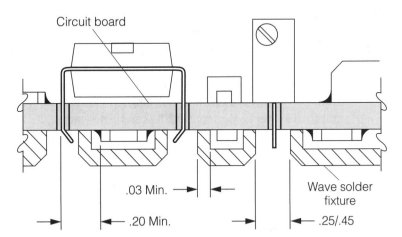

■ 12-3 *When planning for selective wave solder, fixture clearance must be maintained between the lead of the component mounted on the primary side of the board and the surface mount device solder attached to the secondary side.*

The following checklist will act as a guide during evaluation of the assembly:

☐ Solder mask separation of land pattern and via pad.

☐ Correct solder mask clearance around land patterns.

☐ Solder mask not present on SMT land pattern surface.

☐ Passive component spacing of 0.8 mm (0.030 in) minimum, side to side.

☐ Via hole and pad is not located under body of small chips.

☐ Component-to-board edge clearance for machine handling.

☐ Space between components for visual solder inspection.

☐ Orientation of parts clearly identified.

☐ Components aligned after reflow-solder process.

☐ Correct IC lead registration and spacing.

☐ Adequate solder fillet for inspection (and touch-up).

☐ Lead spacing on PIH devices correct for machine insertion.

☐ Nonplated tooling holes provided for machine fixtures.

☐ Three global fiducial (optical) targets for vision alignment.

☐ Local fiducial targets provided for fine-pitch device placement.

Design for test

One of the most critical elements in circuit board design is providing for automated testing of the final assembly. When in-circuit testing has been defined for the assembly, test probe contacts must be provided. The identification of the malfunctioning device is difficult and costly with bench-top analyzing equipment. By using In-Circuit-Test systems, each part value can be measured, IC device functions exercised, and the entire assembly screened for solder opens and shorts. If test probe contact patterns are not accessible at each common junction of the circuit network, measuring or exercising the components individually in the circuit will not be possible.

General considerations for in-circuit test automation

When reviewing the assembly for testability, verify the following requirements:

☐ One test probe contact node per net for in-circuit test.

☐ Test probe contact spacing of 2.0 mm (0.080 in) minimum.

☐ Probe contact area of 0.9 – 1.0 mm (0.035 – 0.040 in) diameter.

☐ Assembly is testable from one side.

☐ Probe body-to-component body clearance of 0.8 mm (0.030 in) minimum.

Do not rely on the component lead or land pattern for test probe contact. The pressure of the spring-loaded probe pin can add just enough pressure to a substandard solder connection to mask a solder defect, then when the pressure is released, the component lead is allowed to open and electrical continuity is broken.

General design recommendations

Following the recommendations outlined and illustrated in the preceding chapters, the designer is prepared with the fundamentals necessary to produce a successful SMT assembly. In general, a few subtle elements of good design practice should be reviewed:

☐ Use process-compatible SMT land pattern design.

☐ Retain consistent orientation on components when possible.

☐ Mount polarized devices in the same direction.

☐ Allow reasonable component density and even distribution.

☐ Select SMT devices in standard configurations.

☐ Choose parts that will have multiple sources.

☐ Design the circuit board to control excessive costs.

Issues related to fine pitch

Fine-pitch assembly typically represents an advanced packaging and manufacturing concept. Component lead density and complexity is far greater on the fine-pitch device than for the majority of commercial products, and if assembled in high volume, several issues must be addressed before an automated process can be developed.

For example, the fine-pitch components developed for custom application-specific integrated circuits (ASIC) have broad mechanical tolerance limits and increased lead density, which require land pattern refinement before process efficient yields are achieved, see Fig. 12-4.

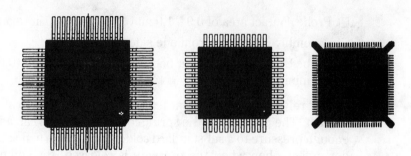

■ **12-4** *QFP devices have a greater lead count in a smaller area than the PLCC family. Spacing between leads often varies from one manufacturer to another.*

Test automation concerns

Of particular concern on high-density fine-pitch assemblies is the accessibility to all of the circuit "nets" or common connections between devices for probing with automated test systems. The in-circuit test systems are programmed to automatically measure function of each device on the completed assembly, locate improperly placed or damaged components, and identify solder process defects. If the design does not meet the "one node per net" criteria, assembly processes, defect detection, and reliable electrical testing might be compromised. The process defects include both bridging of solder between the leads of devices and open connections between the device lead and the boards land pattern as illustrated in Fig. 12-5.

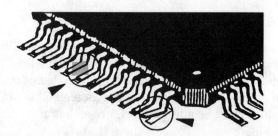

■ **12-5** *Process defects include both bridging of solder between the leads of devices (left) and open connections between the device lead and the board's land pattern (right).*

Process control and inspection

Quality inspectors follow established procedures and industry standards to measure specific device-mounting quality. These standards have established criteria for device alignment, solder fillet quality, and they identify solder defects that relate to process control. All inspection methods are typically based on visual examination using low-power magnifiers. Solder joint quality, for example, can be measured against a criteria defined in both IPC-A-620 for overall workmanship and the National Solder Standard, ANSI/J-STD-001, which defines materials and verification of the solder joint quality level or class. During this procedure the inspector is responsible for determining evidence of wetting to confirm that an acceptable solder fillet profile is provided.

Assembly process development

Process development for fine-pitch surface mount assembly is not routine. Although mathematical modeling has proved to be a valuable tool in developing this process, it is often through practical experience that engineers refine the overall process. Tailoring proven methods developed for other assemblies for soldering has helped companies meet quality criteria while yielding a reliable finished product. Because each assembly has its own character, the process engineering specialist will develop a detailed *Manufacturing Process Instruction* (MPI) document. Each product is unique, however, and generally requires special fixturing dedicated to individual assembly type.

For example, if surface mount devices are to be reflow solder attached to both the primary and secondary side of the board, two stencil fixtures must be provided and special vacuum fixtures might be needed to hold the board during the solder application process.

When fine-pitch devices are to be attached on both sides of a PC board and some devices are exposed to the in-line reflow process twice, then solder migration on the previously attached device might allow the leads to desolder or de-wet. To avoid exposure of devices attached during the first-pass side during the second reflow process, a masking fixture is often designed. These fixtures can be machined from a rigid dielectric material (generally FR-4 or high-temperature polyimide) to shield the previously soldered devices from exposure to the full temperature of the oven. An additional holding fixture might be necessary if wave-solder attachment of the leaded through-hole components is necessary. The wave-solder fixture can shield all surface mount devices already attached using reflow-solder processes from exposure to the wave while allowing limited or selective access to the leads to be soldered.

Assembly processes for SMT includes extensive monitoring through visual inspection at each operation. As an example measurement of solder paste volume with a laser scanning microscope on each board is recommended followed by additional periodic sampling of the product during other stages of the assembly sequence.

Requirements for process development

Processing even a minimum volume of boards can give the engineer the opportunity to test and refine all process variables that might be unique to each assembly type. To ensure that the process development phase of the program has adequate materials for process refinement, provide one or two sets of boards and materials for experimental use.

Understanding that some devices can represent a significant value, and specific components might have limited quantity, it is advised that the experimental process be conducted using the nonfunctional or "dummy" parts during the early stage of process development. Before and during the initial manufacturing start-up, the MPI might continue to be modified or refined as necessary, however, when the development phase of the program is completed, the document should ensure consistent process uniformity and repeatability. Figure 12-6 illustrates design layout and land pattern recommendations that will contribute to both manufacturability and reliability of the fine-pitch product.

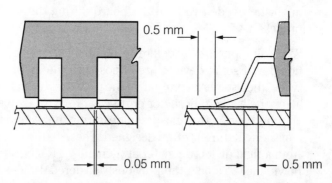

■ **12-6** *Design layout and land pattern recommendations that will contribute to both manufacturability and reliability of the fine-pitch product.*

EIAJ/JEDEC Shrink Quad Flat Pack (SQFP)

Fine-pitch QFP devices are available with lead pitch both greater and less than 0.025 in. For example, QFP devices with the 1.0 mm (0.040 in) and 0.8 mm (0.031 in) lead spacing is available for custom and semi-custom IC applications but newer, high- performance devices are more likely going to be furnished in lead pitch of 0.5 mm (0.2 in) or less. These families are developed using the international "metric" (millimeter) rather than in inches, providing a more precise physical tolerance for the smaller features.

One of the more popular families of fine-pitch products is a package identified as SQFP or Shrink Quad Flat Pack. Both a square and a rectangular package are offered for the SQFP and the standard pitch options includes 0.5 mm, 0.4 mm and 0.3 mm.

Square SQFP Device

Typical of the QFP, the gull wing leads extend away from each of the four edges. As noted, three lead pitch options are offered and the package size range for the square SQFP is from 5.0 mm (0.197 in) through 44.0 mm (1.74 in). As an example, Fig. 12-7 illustrates one of the package configurations available. The maximum I/O for the 44.0 mm (1.74 in) square package with 0.3 mm(0.012 in) pitch is 576 total leads. It is expected that with all the choices available in the standard, this low-profile package family will accommodate the silicon chips for current usage, as well as the foreseeable future.

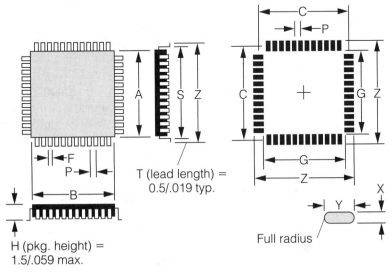

T (lead length) = 0.5/.019 typ.

H (pkg. height) = 1.5/.059 max.

Full radius

■ **12-7** *Low-profile QFP devices introduced by the EIAJ offer a wide range of I/O choices and three lead pitch configurations.*

Rectangular SQFP Devices

Six rectangular packages are defined in the EIAJ/JEDEC standard. The rectangular SQFP defined in Fig. 12-8 has the same lead space options as the square version, but the lead counts will reflect the uneven side dimension. Rectangular lead frame design is offered for packaging the nonsymmetrical silicon die shapes.

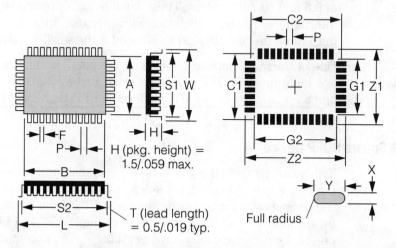

■ **12-8** *The rectangular QFP outline is a more limited family than the square QFP but offers an efficient package alternative for housing the rectangular shaped die.*

While the use of miniature fine-pitch devices has provided layout flexibility, the need to push components even closer together on very complex multilayer substrates has also impacted the cost of the product. Companies adapting fine pitch are attempting to relieve the growing space dilemma by combining the function of several integrated circuits into one or two multi-function custom devices. The customized or application specific ICs might alleviate the PC board grid-lock but higher lead count and finer lead pitch will generally compel the designer to use more circuit layers on the PC board, thus increasing PC board fabrication cost.

Thin Small Outline Package (TSOP)

Like the SOIC packages, the TSOP has two rows of gull wing formed leads on opposing sides. Except for the gull wing lead shape, TSOP and SOIC packages are physically different however. The leads of the SOIC and TSOP Type II are on the long edge of the rectangular device. The TSOP Type I family is designed for lead exit at the end of the rectangular shape. The SO and SOP package has a body height above 1.5 mm (0.060 in) while the TSOP devices are typically a lower profile 1.27 mm (0.050 in) in height. Lead spacing of the TSOP is also finer than the SO/SOP device (1.27 mm (0.050 in) pitch). Four TSOP Type I package sizes are established as standard with lead pitch and I/O configuration to accommodate a variety of device applications. Figure 12-9 describes the nominal dimensions for the TSOP Type I device family.

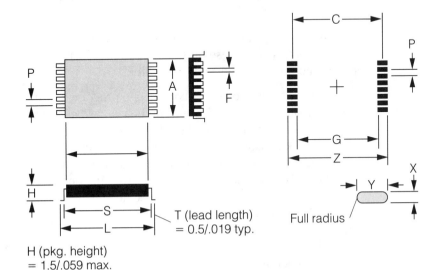

H (pkg. height)
= 1.5/.059 max.

■ **12-9** *Thin Small Outline Package outlines offer a choice of lead pitch and accommodate two distinct configurations. The TSOP Type I furnishes the lead at the end of the device while the Type II version locates the leads on the side of the package.*

Developed to meet the low-profile needs of the PCMCIA and products having a restricted overall height, the TSOP is widely used throughout the industry. Having all leads at opposite ends of the package accommodates routing of the circuit and has proven ideal for memory products. The smart card standards limit the overall thickness of the PCMCIA assembly. The thin devices, generally attached to a 0.5 mm (0.020 in) thick substrate, can stay within the limits established and even allow TSOP attachment on both primary and secondary sides of the substrate. The PCMCIA meets the physical standard established by the Personal Computer Memory Card Industry Association.

Process monitoring for two-sided SMT

Recognizing that the fine-pitch design is a far more complex assembly than the average surface mount product, alternative process and test methods must be considered to ensure manufacturing efficiency and yield. The component density factor and a high percentage of fine-pitch devices on advanced products adds a unique level of difficulty to assembly processes and inspection as well.

Because the component density can be equal on both primary and secondary sides of the assembly, conventional electrical testing for process monitoring might not be possible.

The in-circuit test systems are fixtured and programmed specifically for each assembly type and when the board has been designed for testability, the system will efficiently detect both process (solder) defects, workmanship quality problems, and component failure. Even though most component malfunctions can be detected through in-circuit and diagnostic capability built into a functional test system, there is considerable concern that specific solder process defects will be difficult to identify. On high-density two-sided SMT assemblies when in-circuit test is not practical, X-ray inspection might offer a solution. The X-ray procedure can be performed after solder processing each assembly side to confirm both consistency of the solder volume at each attachment point and to observe any change in the solder joint profile that might occur during the second reflow process.

Surface mount technology, although mature, continues to evolve. The design and assembly processes are interrelated to a higher degree than the PIH products of past years. Design engineers, CAD specialists, and manufacturing disciplines must work together in planning each phase of the surface mount product. More than an evolution, surface mount technology has proved to be a revolution within the ever-expanding electronic manufacturing industry.

In addition to adhering to proven design-related factors, components and processes are being constantly refined. Fine-pitch components developed for both general-purpose and application-specific integrated circuits (ASIC) have increased lead density, and this in turn will require the development of unique land patterns, interconnection methods, improved solder technology, and innovative design solutions.

252

Index

Illustrations are in **boldface**.

254

255

H

heat seals, 96-97, **97**
high reliability electronic products (Class 3/Level C), 3, 4-6, **5**

I

in-circuit testing, 162-163, 244-245
inductors, **15**, 26, **27**
    land pattern geometry, 55-56
inspection, 246-247
Institute for Interconnecting & Packaging Electronic Circuits (IPC), 4
integrated circuits (ICs), 30-31, **31**
    ball grid array (BGA) ICs, 42-44
    ceramic column grid array (CGA) ICs, 44
    die grid array (DGA) ICs, 44-45
    land pattern geometry, plastic diode packages, 55
    plastic lead chip carrier (PLCC), 30, 31, 35-37, **36**, **37**, **38**
    quad flat pack (QFP), 31, 39-40, **40**, **41**
    shrink quad flat pack (SQFP), 40-42, **41**
    small outline IC (SOIC), 30, 31-34, **32**, **33**
    small outline J lead (SOJ), 30, 37
    thin small outline package (TSOP), 34-35, **35**
Invar, copper-clad, 201, 202-206, **202**, **203**, **204**
IPC-D-279, 6
IPC-SM-782 standard, 6
IPC-SM-785 test standards, 8

J

Joint Electronic Device Engineering Council (JEDEC) standards, 1, 13, 14

L

laminated substrate materials, 174-175
land pattern geometry, 11, 47-77
    ball grid array (BGA), 71
    bridging, 52, **53**
    CAD/CAM design requirements, 72-73, **73**
    capacitors, 48-55, **49**
    ceramic devices, 69-70, **70**, **71**
    computer aided design (CAD), 47
    coverlay, 198
    coverlay openings, 142-143, **143**
    die grid array (DGA), 71-72

epoxy-attached components and wave-soldering, 53, **54**
    fillet and height requirements/restrictions, 51, **51**
    fine-pitch devices, 151, **151**, **152**, 155, **156**, 158-159, **159**
    flexible circuits, 136-137, **137**, 142-143, **143**
    inductors, 55-56
    MELF components, 54-55, **54**
    orientation of devices, 74-76, **74**, **75**
    passive components, 48-55
    physical size limits/dimensions, 49, **50**
    planning for land pattern geometry, 47-48
    plastic diode land patterns, 55
    plastic leaded chip carrier (PLCC) ICs, 64-67, **65**, **66**, **67**, **68**
    polarity of devices, 74-75
    quad flat pack (QFP) ICs, 67-68, **69**
    reflow-solder processing, self-centering, 49, **50**, **51**
    resistors, 48-55, **49**
    rigid circuits, 108, **109**, 110-112, **111**
    small outline ICs (SOIC), 59-62, **60**, **61**, **63**
    small outline J lead (SOJ) ICs, 63-64, **64**
    small outline packages (SOP), EIAJ standards, 61-62, **63**
    SOT-143 transistor package, 59, **59**
    SOT-23 transistor package, 56-57, **57**, **58**
    SOT-89 transistor package, 57-58, **57**
    spacing, 92-95, **92**, **93**, **94**, **95**
    stencils, 77, 154, **154**, **155**, 224-225, **225**
    tabs, breakaway tabs, 178, 179, **179**
    tanatalum capacitor land pattern, 55-56, **56**
    thin small outline packages (TSOP), 62-63
    transistors, 56-59, **57**, **58**, **59**
    tube- or tray-carrier devices, 76
    wave-soldering
        passive component land patterns, 51-53, **52**, **53**, **54**
        rigid circuit land patterns, 129-132, **131**
    X- and Y-coordinates for automated manufacturing, 76, **76**
layout, 11
    flexible circuits (*see* flexible circuit layout)
    rigid circuits (*see* rigid circuit layout)
leadless ceramic chip carrier (LCC) (*see* ceramic devices)

functional cluster testing, 163-164
in-circuit testing, 162-163, 244-245
IPC-SM-785 test standards, 8
planning for automated testing, 161-162
test points, masks and cleaning concerns, 231-232, **232**
test probe design options, 164, **165**
test-fixture preparation, 171
two-sided boards, 168-169, **169**
viscosity testing, 224
thermal coefficient of expansion (TCE)
ceramic devices, 70
substrate materials, 176
thin small outline package (TSOP), 34-35, **35**
land pattern geometry, 62-63
through-hole assembly, **9**
tombstoning, 86, 118
trace-to-contact guidelines
conductor trace routing, 119
flexible circuits, 139, **139**, 142
rigid circuits, 117-118, **118**, 119
spacing, 85-88, **86**, **87**, **88**, 117-118, **118**, 119
substrates, 183-184, **184**
transistors, **15**, 27-29, **29**, 29-30, **30**
array packaging, 29-30, **30**
bipolar transistors, 28
field effect (FET), 28
land pattern geometry, 56-59, **57**, **58**, **59**
MELF style, 29
SOT-143 transistor package, land pattern geometry, 59, **59**
SOT-23 transistor package, land pattern geometry, 56-57, **57**, **58**
SOT-89 transistor package, land pattern geometry, 57-58, **57**
two-sided boards, 208-209, **208**, **209**, 248-249, **249**
flexible circuits, conductor routing, 140, **141**

spacing, 83-84, **85**, 95
substrates, 180-181, 187, 194-195
testing, 168-169, **169**

U

use categories: Class 1, 2, 3 products), 3, 6, **7**, 8

V

vapor-phase reflow-soldering, 210-211, **211**
via hole/pad connections
diameter and hole spacing, leaded devices, 122, **123**
flexible circuits, hole size to pad, 141
plated through-holes, 184-186, **185**, **186**
rigid circuits, 110-112, **112**, 114-116, **115**, **116**, 119, **119**
substrates, 181-183, **182**, **183**
voids, 91

W

wave-soldering, 8, 9, 209-210, **209**
bridging defects, 52, **53**
component-to-edge clearances, 132-133, **132**, **133**
double reflow-wave soldering, 134
epoxy-attached components and wave-soldering, 53, **54**
land pattern geometry, 51-53, **52**, **53**, **54**, 129-132, **131**
rigid circuits, 128-132, **130**, **131**, 133-134
spacing, 91-95, **92**, **93**, **94**, **95**

X

X7R stable capacitor, 23-24

Y

Y5V (Z5U) general purpose capacitor, 24

261